66 Springer Series in Solid-State Sciences
Edited by Peter Fulde

Springer Series in Solid-State Sciences

Editors: M. Cardona   P. Fulde   K. von Klitzing   H.-J. Queisser

Peter Brüesch

Phonons: Theory and Experiments III

Phenomena Related to Phonons

With contributions by
J. Bernasconi, U. T. Höchli, and L. Pietronero

With 110 Figures

Springer-Verlag Berlin Heidelberg New York
London Paris Tokyo

Professor Dr. Peter Brüesch

Brown Boveri Research Center, CH-5045 Baden-Dättwil, and
Ecole Polytechnique Fédérale de Lausanne, CH-1015 Lausanne, Switzerland

Contributors:

Dr. J. Bernasconi

Brown Boveri Research Center
CH-5405 Baden-Dättwil, Switzerland

Dr. U. T. Höchli

IBM Research Center
CH-8803 Rüschlikon, Switzerland

Professor L. Pietronero

Solid State Physics Laboratory
University of Groningen, Melkweg 1
NL-9718 EP Groningen, The Netherlands

Series Editors:

Professor Dr., Dr. h. c. Manuel Cardona
Professor Dr., Dr. h. c. Peter Fulde
Professor Dr. Klaus von Klitzing
Professor Dr. Hans-Joachim Queisser

Max-Planck-Institut für Festkörperforschung, Heisenbergstrasse 1
D-7000 Stuttgart 80, Fed. Rep. of Germany

ISBN 978-3-642-52273-4   ISBN 978-3-642-52271-0 (eBook)
DOI 10.1007/978-3-642-52271-0

Library of Congress Cataloging-in-Publication Data. (Revised for vol. 3) Brüesch, Peter, 1934- Phonons, theory and experiments. (Springer series in solid-state sciences ; 34-) Includes bibliographical references and index. Contents: Lattice dynamics and models of interatomic forces. 1. Phonons. I. Series: Springer series in solid-state science ; 34, etc. I. Title. QC176.8.P5B78 1982 530.4'1 81-21424

© Springer-Verlag Berlin Heidelberg 1987
Softcover reprint of the hardcover 1st edition 1987

2153/3150-543210

*Dedicated to my wife
and my parents*

Preface

The first volume of this treatment, *Phonons: Theory and Experiments I*, was devoted to the basic concepts of the physics of phonons and to a study of models for interatomic forces. The second volume, *Phonons: Theory and Experiments II*, contains a study of experimental techniques and the interpretation of experimental results. In the present third volume we treat a number of phenomena which are directly related to phonons.

The aim of this book is to bridge the gap between theory and experiment. An attempt has been made to present the descriptive as well as the analytical aspects of the topics. Although emphasis is placed on the role of phonons in the different topics, most chapters also contain a general introduction into the specific subject. The book is addressed to experimentalists and to theoreticians working in the vast field of dynamical properties of solids. It will also prove useful to graduate students starting research in this or related fields.

The choice of the topics treated was partly determined by the author's own activity in these areas. This is particularly the case for the chapters dealing with phonons in one-dimensional metals, disordered systems, superionic conductors and certain newer aspects of ferroelectricity and melting.

I am very grateful to my colleagues J. Bernasconi, U.T. Höchli and L. Pietronero who wrote some of the chapters or sections in this volume:

J. Bernasconi: Phonons in Disordered Systems (Chap. 6)

U.T. Höchli: Range of Validity of the Landau-Devonshire Theory (Sect. 3.5.4); Quantum Ferroelectricity (Sect. 3.7); Disordered Polar Systems (Sect. 3.8)

L. Pietronero: The Electronic Susceptibility of 1-D Metals (Sect. 5.3); The Electron-Phonon Hamiltonian of 1-D Metals (Sect. 5.4); Melting (Chap. 8)

The author is indebted to BBC Brown, Boveri & Company Limited, Baden, Switzerland, for giving him the opportunity to carry out this work. The BBC Research Center provided the necessary scientific atmosphere to accomplish this goal. As in the first and the second volumes, some chapters of this third volume grew out of lectures the author gave at the "Ecole Polytechnique Fédérale de Lausanne" (EPFL) during the years 1975–1985 for graduate students in experimental physics in their last year of study. I am grateful to BBC and to EPFL for giving me the chance to prepare parts of this book by lecturing.

   The entire manuscript was strongly influenced by the detailed criticism and valuable suggestions of my colleagues Drs. Th. Baumann, J. Bernasconi, E. Cartier, T. Hibma, P. Pfluger, W.R. Schneider, H.J. Wiesmann and H.R. Zeller from the BBC Research Center, and by Dr. W. Bührer from the Institut für Reaktortechnik ETHZ in Würenlingen, Switzerland. My debt to them is great. I am also grateful to Professor Ph. Choquard from EPFL for many interesting discussions in Lausanne and to Professor P. Erdös from the University of Lausanne for his critical reading of the chapter about thermal conductivity.

   I wish to express most grateful thanks to Mrs. M. Zamfirescu from the BBC Research Center for the very skilful drawing of over 100 figures; without her immense work the goal of presenting a well-illustrated book could not have been achieved. The author is also indebted to the late Mrs. E. Knotz, to Mrs. N. Bingham and to Mrs. E. Martens for their never-ending patience in typing the manuscript. Finally, I am grateful to Professor P. Fulde for valuable suggestions and to Dr. H. Lotsch, Springer-Verlag, for good cooperation.

Baden, November 1986   *Peter Brüesch*

Contents

1. Introduction

The first volume of this treatment, *Phonons: Theory and Experiments I* [1.1], was devoted to the basic concepts of the physics of phonons and to a study of models for interatomic forces. In the second volume *Phonons: Theory and Experiments II* [1.2], a thorough study has been given to experimental techniques and the interpretation of experimental results. In the present third volume we treat a number of phenomena which are directly related to phonons.

1.1 General Remarks

After a certain period of active research each branch of science reaches a state of maturity which is characterized by the fact that the basic concepts within the field itself are established. In this stage further progress is expected to arise through interactions with other branches of research. It has been recognized that phonons play a crucial role in many important phenomena in classical and modern solid state physics. Figure 1.1 illustrates the interaction of phonons with some of these phenomena. Phonons

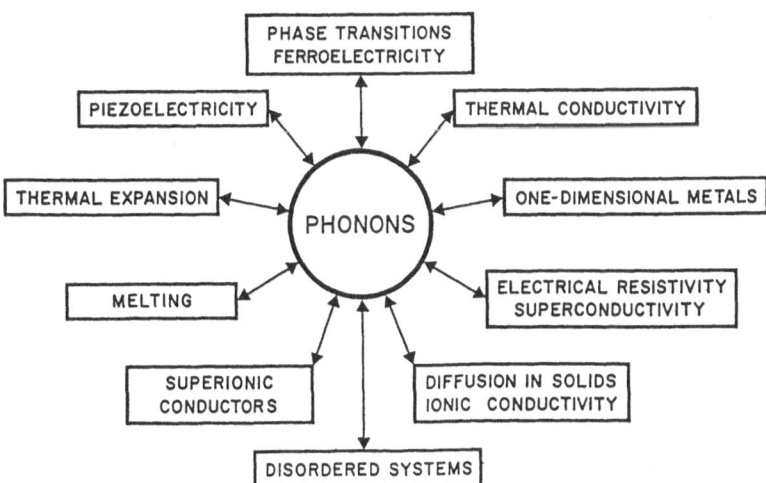

Fig. 1.1. Interaction of phonons with some branches of solid state physics. In the topics indicated phonons play a crucial role in the microscopic mechanism involved

are of prime importance in such basic phenomena as thermal expansion, piezoelectricity, phase transitions, thermal conductivity, electrical resistivity, superconductivity, diffusion of atoms in solids, ionic conductivity and melting. More recently the role of phonons has been intensively studied in one-dimensional metals and superionic conductors. Another active field of research is concerned with phonons in disordered systems and with surface phonons.

It is the aim of this book to investigate the role of phonons in relation to piezoelectricity, ferroelectricity, thermal conductivity, one-dimensional metals, disordered systems, superionic conductors and melting. Thermal expansion and the lattice dynamics of molecular crystals have been treated in [1.1]. Other topics in which phonons also play an important role are not treated and the reader is referred to the literature. Examples are electrical resistivity [1.3,4], superconductivity [1.5,6], interaction of magnetic ions with phonons [1.7,8] and lattice dynamics of quantum crystals [1.9].

1.2 Piezoelectricity

Chapter 2 is devoted to piezoelectricity. It starts with a macroscopic description based on electro-elastic relations. The phenomenon is then treated on a microscopic basis by using two different methods, the method of homogeneous deformations and the method of long waves. Both methods are illustrated with the help of a one-dimensional diatomic model which mirrors the situation in crystals such as ZnS for deformations parallel to the (111) direction. The results are then applied to piezoelectricity of ZnS crystals.

1.3 Ferroelectricity

In Chap. 3 we first discuss the basic facts about ferroelectricity using the thermodynamic theory of Landau-Devonshire for first- and second order transitions. We then present the general aspects of the lattice dynamics of displacive ferroelectric phase transitions, including the lattice dynamical basis of Devonshire's theory. This is followed by some lattice dynamical models which have been developed for the study of ferroelectric phase transitions. The last sections are devoted to a discussion of soft mode spectroscopy of selected ferroelectrics, to quantum ferroelectricity and to polar disordered systems.

1.4 Thermal Conductivity

Thermal conductivity is treated in Chap. 4. After an introduction and a discussion of the experimental determination of thermal conductivity, the subject is first treated on the basis of the elementary kinetic theory. The formal theory of thermal conductivity which is based on phonons and on the Boltzmann transport equation is presented in Sect. 4.4, while Sect. 4.5 is devoted to a discussion of relaxation times in insulators. Finally a short discussion is given about thermal conductivity in glasses, metals and alloys, and about the phenomena of second sound.

1.5 Phonons in One-Dimensional Metals

In one-dimensional metals the lattice dynamics cannot be separated from the dynamics of the conduction electrons. One-dimensional metals therefore provide simple model systems for the study of the electron-phonon interaction. After an introduction into the field we discuss the basic properties of one-dimensional metals. In Sects. 5.3 and 4 these properties are derived on the basis of the mean field theory by considering the electronic susceptibility and the electron-phonon Hamiltonian of a one-dimensional system. Section 5.5 is devoted to a discussion of important experiments performed on KCP, such as the Peierls distortion and charge density wave (CDW), the Kohn anomaly and the pinned Fröhlich mode. The effects of fluctuations and three-dimensional coupling are discussed in Sect. 5.6.

1.6 Phonons in Disordered Systems

In a perfect crystalline material, the translational invariance of the lattice considerably simplifies a theoretical investigation of the vibrational properties. Interesting real materials, however, are often disordered, and even carefully prepared crystals invariably contain defects or impurities. The corresponding perturbation, or even destruction, of the translational invariance modifies the vibrational properties of the material. Phonons in disordered systems are studied in Chap. 6. In this chapter we restrict ourselves to a brief description and illustration of some important theoretical methods by applying them to simple linear chain systems. The corresponding treatment of more complicated systems is discussed on a rather general level only, and a comparison between theoretical calculations and experimental results is restricted to a few illustrative cases.

1.7 Ion Dynamics in Superionic Conductors

The ion dynamics in superionic conductors (SIC) is treated in Chap. 7. SIC are solid electrolytes which are characterized by an almost liquid-like mobility of one of the ionic species. This is due to the fact that the mobile ions are disordered and distributed over a large number of sites of the host lattice, and can jump with small activation energies from one site to neighbouring empty sites. After an introduction into the subject, the chapter starts with a discussion of ionic interactions and dynamical models in SIC. In Sect. 7.4 the simplest model to describe ionic motion is introduced, the Brownian motion of a particle in a periodic potential. Disorder-induced infrared absorption and Raman scattering is discussed qualitatively in Sect. 7.5, while Sect. 7.6 contains a short discussion of the study of phonons in α-AgI and some other SIC. Finally, we discuss an approximate lattice dynamical calculation of jump frequencies which serves to illustrate the connection between ionic conductivity and lattice dynamics.

1.8 Melting

The last chapter is concerned with melting. The study of melting on a microscopic basis is an old and difficult subject. In Chap. 8 the general features of the melting transition for three-dimensional simple solids are critically discussed. In particular the relations between melting and the instabilities of anharmonic lattice dynamics are treated in some detail. The possibility of *surface melting* is also considered. The theoretical results suggest that there should be a temperature T_s lower than the real melting temperature T_M, at which the free surface of a solid becomes quasi-liquid. Experiments and computer simulations are rather consistent with this picture. In fact, a recent experimental study on lead strongly confirms such a picture.

2. Piezoelectricity

After a general discussion into the physical aspects of piezoelectricity in Sect. 2.1, the following topics are treated: In Sect. 2.2 the macroscopic electroelastic relations are established and discussed. In Sect. 2.3 we consider the static approach to piezoelectricity, i.e. the method of homogeneous deformations. Section 2.4 is devoted to establish a direct relation between piezoelectricity and lattice dynamics based on the "method of long waves". The essential physical aspects are demonstrated by using a simple one-dimensional model. In Sect. 2.5 this model is applied to a treatment of piezoelectricity of crystals with ZnS-structure. Section 2.6 contains a number of problems which provide additional information.

2.1 General Remarks

The piezoelectric effect is the production of electric polarization by application of stress to a crystal. This effect has been discovered by the brothers Curie in 1881 [2.1]. They found piezoelectricity in zinc blende, sodium chlorate, boracite, tourmaline, quartz, Rochelle salt and in a number of other compounds. The inverse piezoelectric effect, that is the change in dimensions of the crystal on the application of a voltage, was predicted theoretically on thermodynamic principles and was first verified by the Curies the following year. Although some of the relations between piezoelectricity and crystal structure were established by the Curies, it was most rigorously determined by *Voigt* in 1884 [2.2]. By combining the elements of symmetry of elastic tensors and of electric vectors with the geometrical symmetry elements of crystals he made clear in which of the 32 crystal classes piezoelectric effects may exist, and for each class he showed which of the possible 18 piezoelectric coefficients may have values different from zero. An excellent treatment of piezoelectricity in relation with crystal symmetry is found in the book of *Nye* [2.3]. For comprehensive treatments of piezoelectricity the reader is referred to the books of *Cady* [2.4] and *Warren* [2.5].

The phenomenon of *electrostriction* should not be confused with the inverse piezoelectric effect. In the latter the deformation is linear in the applied field, while electrostriction is independent on the direction of the field and proportional to the square of the field.

In general, ionic or partly ionic crystals are piezoelectric. The crystal when stressed develops a polarization due to the relative displacements of

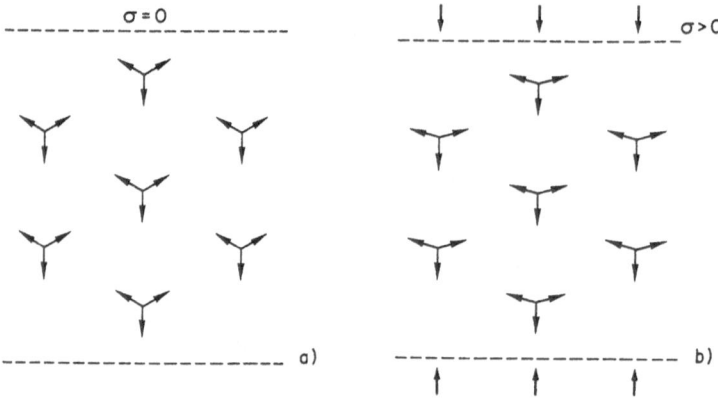

Fig. 2.1. (a) The unstressed crystal contains planar group of complex ions, such as for example $(CO_3)^{-2}$. The arrows represent dipole moments and the sum of the three dipole moments of each complex ion with trigonal symmetry is zero. (b) The crystal when stressed develops a polarization, because the sum of the dipole moments at each complex ion is no longer zero

positive and negative charges[1]. This is shown schematically in Fig. 2.1. In the absence of an external stress the individual dipole moments cancel each other and no polarization will occur. A stress will change the orientation of the dipoles in such a way that each group of ions acquires a dipole moment and all these dipole moments add up resulting in a non-vanishing polarization P.

There are, however, structures where the application of a stress does not give rise to a macroscopic polarization, because the individual dipole moments cancel each other. This is illustrated in Fig. 2.2a, b in which each complex ion is located at a centre of inversion which is not destroyed by the application of a stress. The alkali halides NaCl- and CsCl-structure [Ref. 2.6, Figs. 4.4 and 3.4] are examples where each ion is located at a centre of inversion. A macroscopic polarization will not occur either if the centre of inversion does not coincide with an ion or the centre of molecule as shown schematically in Fig. 2.2c, d. *Hence, a crystal with a centre of symmetry cannot be piezoelectric.*

The zincblende structure (Fig. 2.3) is the simplest to show the piezoelectric effect. This structure is characterized by the fact that planes perpendicular to a body diagonal, the (111)-direction for example, are occupied

---

[1] A permanent polarization may already exist before application of a stress or an electric field; this is a ferroelectric crystal (Chap. 3). When a stress is applied to a ferroelectric crystal the polarization changes; therefore, a ferroelectric crystal is always piezoelectric, but the converse is not true.

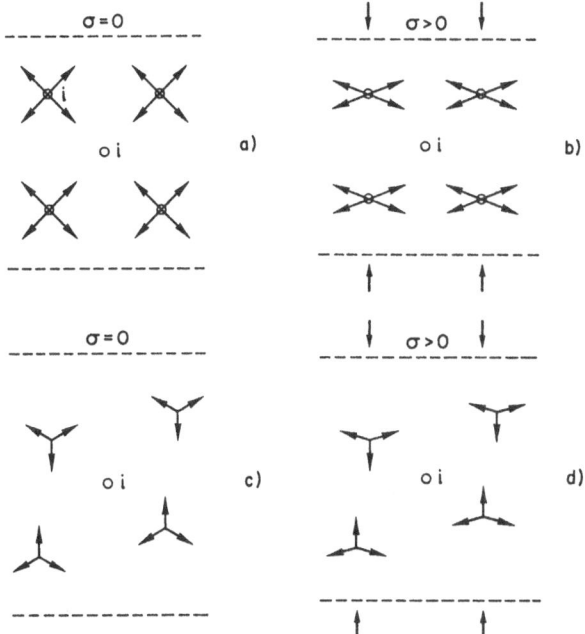

Fig. 2.2. (a) In the unstressed crystal each molecule with polar bonds indicated by arrows is located at a centre of symmetry and the sum of the dipole moments vanish. (b) In the stressed crystal the molecules are deformed but remain on a centre of symmetry and consequently the polarization P is still zero. (c) The complex ions in the unstrained crystal are related by a centre of symmetry and $P = 0$. (d) In the stressed crystal the complex ions are deformed but they are still related by a centre of symmetry, and consequently $P = 0$

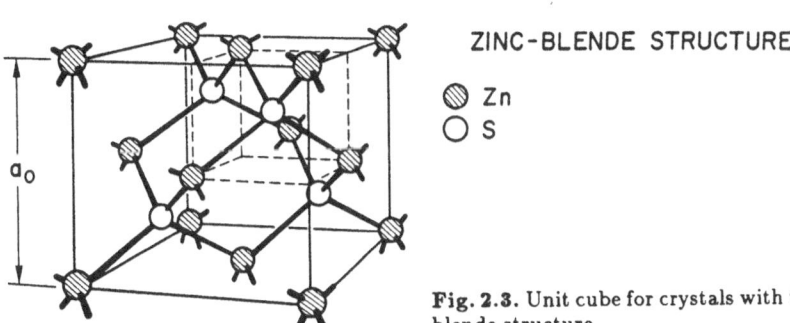

ZINC-BLENDE STRUCTURE

⊘ Zn
◯ S

Fig. 2.3. Unit cube for crystals with the zinc-blende structure

by one type of ions only, either Zn or S ions, and that the distances between neighbouring planes are quite different; this situation is shown schematically in Fig. 2.5. If a stress is applied parallel to a body diagonal of the cube, the sublattice of the Zn ions is rigidly displaced relative the sublattice of the S ions in this direction; the relative displacements of the two sublattices gives

7

rise to a macroscopic polarisation. In Sects. 2.3 and 4 we shall discuss the piezoelectric effect in the ZnS structure on the basis of simple models.

The reader might be tempted to assume that piezoelectricity occurs only in ionic or partly ionic crystals or in crystals which contain at least two different atoms in the unit cell. This is not the case. Important counter examples are elements (trigonal selenium and tellurium) shown in Fig. 2.4 [2.7–9]. The atoms form long spirals with trigonal symmetry, and these spirals are arranged in a hexagonal pattern, so that the Bravais lattice is hexagonal. If a stress is applied parallel to the chain axis a very strong polarization is induced. Se and Te are examples of *piezoelectric semiconductors*. How can a macroscopic polarization occur if a stress is applied to such an elemental crystal? It is certainly necessary that the strain causes a relative displacement of positive and negative charges giving rise to a macroscopic polarization. In the case of ZnS, for example, the positive and negative charges which are displaced relative to each other are the effective ionic charges $\pm e$ residing on the Zn and S ions, respectively. In elemental Se and Te such a simple picture does not apply because no effective ionic charges in this sense exist. The strong covalant bonds within a chain can be described by a bond charge model [Ref. 2.6, Sect. 4.4] or possibly by a shell model [Ref. 2.6, Sect. 4.3]. If a stress is applied to the crystal it is possible that relative displacements of the bond charges against the "ionic" charges, or alternatively, of the shell charges relative to the core charges will occur. This is probably the origine of the observed strong piezoelectricity in Se and Te; to the authors knowledge no attempt has been made to calculate the piezoelectric constants of Se and Te on the basis of a lattice dynamical model. At this point it should be mentioned that trigonal Se and Te show strong infrared absorption [2.10–15]. The origin of this infrared absorption is also due to the existance of dynamic effective charges which depend on the mode [2.12] and which can be different from zero even if the nominal charge $ze = 0$ [Ref. 2.6, Sect. 4.3.1]. By the same mechanism a long-wavelength acoustic mode which describes a strained state of the crystal can, in principle, produce a piezoelectric charge in crystals such as

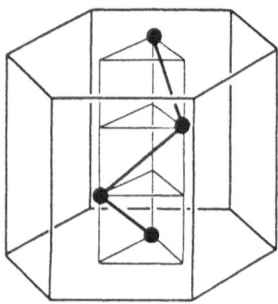

Fig. 2.4. Crystal structure of trigonal Se and Te [2.12]

Se and Te, but not, for instance in diamond which possesses a centre of symmetry.

Piezoelectricity is important for many applications, such as transducers and materials suitable for wave generation. If an ac voltage is applied to a piezoelectric material, mechanical vibrations are generated having the same frequency as that of the voltage; if the frequency of the voltage is the same as a self-frequency of the elastic medium, very intense elastic vibrations can be excited. It should also be mentioned that piezoelectricity is not limited to single crystals. Piezoelectricity has also been observed in polymers and in ceramic materials [2.16]. The most famous piezoelectric polymer is polyvinylidene fluoride which exhibits piezoelectric coefficients as large as 6.7×10^{-12} C/N. This is comparable to quartz and about ten times greater than the piezoelectric coefficients exhibited by other polymers. The origin of piezoelectricity in polymers is, however, still a matter of controversy.

In the following we shall confine ourselves to piezoelectricity in single crystals and we shall emphasize the relation between lattice dynamics and piezoelectricity. Since a rigorous lattice dynamical approach to piezoelectricity is rather involved [2.17,18] we shall demonstrate and discuss the essential physical aspects on the basis of a simple one-dimensional model, which can, however, be directly related to piezoelectricity in crystals with ZnS structure or, in a certain approximation, also with crystals with wurtzite structure. In Sect. 2.2 we start with a discussion of the macroscopic relations describing piezoelectricity. Section 2.3 is devoted to the static approach to piezoelectricity which is based on homogeneous deformations [2.19–21]. In Sect. 2.4 we establish the relation between lattice dynamics and piezoelectricity; this is known as the method of long waves [2.17]. Finally, in Sect. 2.5 we discuss the results obtained for the one-dimensional model and relate it with the piezoelectric effect observed in ZnS.

2.2 Macroscopic Electroelastic Relations

If a stress is applied to certain crystals they develop an electric polarization whose magnitude is proportional to the applied stress. this is known as the *direct piezoelectric effect.* For example, if a uniaxial tensile stress σ is applied along one of the diad axes of a quartz crystal, the magnitude of the electric moment per unit volume, i.e., the polarization P, is proportional to the applied stress,

$$P = d\sigma \ , \tag{2.1}$$

where d is a constant, called the *piezoelectric modulus.* In general, a state of stress is specified by a second-rank tensor with 9 components [Ref. 2.6, Sect. 3.6], while the polarization of the crystal, being a vector, is specified by three components. It is found that when a general stress with components $\sigma_{\alpha\beta}$ ($\alpha, \beta = x, y, z$) acts on a piezoelectric crystal, each component of the polarization, P_α, is linearly related to all the components of σ. We accordingly write

$$P_\alpha = \sum_{\beta\gamma} d_{\alpha\beta\gamma} \sigma_{\beta\gamma} \; . \tag{2.2}$$

The 27 piezoelectric moduli $d_{\alpha\beta\gamma}$ form an example of a third-rank tensor [2.3].

When an electric field is applied in a piezoelectric crystal the shape of the crystal changes slightly. This is known as the *converse piezoelectric effect*. It is found that there is a linear relation between the components E_α of the electric field $\boldsymbol{E}$ within the crystal and the components $\varepsilon_{\beta\gamma}$ of the deformation tensor ε which describe the change of shape [Ref. 2.6, Sect. 3.6]. Moreover, it can be shown on the basis of thermodynamic reasoning [2.3] that the *coefficients connecting the field and the deformation in the converse effect are the same as those connecting the stess and the polarization in the direct effect*. The converse effect is therefore written in the form

$$\varepsilon_{\beta\gamma} = \sum_{\alpha} d_{\alpha\beta\gamma} E_\alpha \; . \tag{2.3}$$

Since $\varepsilon_{\beta\gamma} = \varepsilon_{\gamma\beta}$ [Ref. 2.6, Sect. 3.6] it follows that $d_{\alpha\beta\gamma}$ is symmetrical in γ and β:

$$d_{\alpha\beta\gamma} = d_{\alpha\gamma\beta} \; . \tag{2.4}$$

There are therefore 18 independent piezoelectric moduli in the general case. The number of independent piezoelectric moduli is further reduced by crystal symmetry [2.3]. In the cubic ZnS-structure, for example, there is only one independent piezoelectric modulus, namely $d_{123} = d_{231} = d_{312}$, and if a centre of inversion is present, all piezoelectric moduli vanish as we have seen in Sect. 2.1.

Equation (2.2) gives the polarization induced by an external stress, in the absence of a macroscopic field, while (2.3) gives the deformation induced by a macroscopic field, in the absence of an external stress. In the general case a crystal will be under the simultaneous influence of stress and macroscopic field, thus P and ε will depend on both, σ and E : $P = P(\sigma, E)$ and $\varepsilon = \varepsilon(\sigma, E)$, and we can write[2]

$$dP = \left(\frac{\partial P}{d\sigma}\right)_E d\sigma + \left(\frac{dP}{dE}\right)_\sigma dE \quad \text{and} \tag{2.5}$$

$$d\varepsilon = \left(\frac{\partial \varepsilon}{\partial \sigma}\right)_E d\sigma + \left(\frac{d\varepsilon}{dE}\right)_\sigma dE \; . \tag{2.6}$$

---

[2] We do not consider here the temperature dependence of P and ε.

Written out explicitly and in integral form we have

$$P_\alpha = \sum_{\beta\gamma} d_{\alpha\beta\gamma}\sigma_{\beta\gamma} + \sum_\beta \chi^f_{\alpha\beta}E_\beta \ , \tag{2.7}$$

$$\varepsilon_{\beta\gamma} = \sum_{\lambda\mu} S_{\beta\gamma,\lambda\mu}\sigma_{\lambda\mu} + \sum_\alpha d_{\alpha\beta\gamma}E_\alpha \ . \tag{2.8}$$

In (2.7) the first term is the contribution to the polarization by any stress that may be imposed from outside; the second term is the contribution which the applied field makes to te polarization in an unstressed (free) crystal. In (2.8) the first term gives the deformation due to an applied stress if $\boldsymbol{E} = 0$ and the second term is the deformation due to the piezoelectric effect. Note that the coefficients of the first term in (2.7) and the second term in (2.8) are the same as discussed above. In a non-piezoelectric crystal these two terms varnish; the polarization is then entirely due to the macroscopic field and proportional to the susceptibility of the free crystal, χ^f, while the deformation is completely determined by the applied stress. χ^f is called the free susceptibility because σ is held constant (corresponding to a crystal under constant pressure). In the absence of a macroscopic field $\boldsymbol{E}$, (2.8) reduces to Hook's law discussed in [Ref. 2.6, Sect. 3.6] and the coefficients $S_{\beta\gamma,\lambda\mu}$ are the elastic compliance constants.

Alternatively, we can consider the polarization P and the stress σ as a function of the deformation ε and the field E : $P = P(\varepsilon, E)$ and $\sigma = \sigma(\varepsilon, E)$. In this case the macroscopic relations are written in the form

$$P_\alpha = \sum_{\beta\gamma} e_{\alpha\beta\gamma}\varepsilon_{\beta\gamma} + \sum_\beta \chi^c_{\alpha\beta}E_\beta \ , \tag{2.9}$$

$$\sigma_{\alpha\beta} = \sum_{\gamma\lambda} C_{\alpha\beta,\gamma\lambda}\varepsilon_{\gamma\lambda} - \sum_\gamma e_{\gamma\alpha\beta}E_\gamma \ . \tag{2.10}$$

The coefficients $e_{\alpha\beta\gamma}$ are called the *piezoelectric constants*. Note that apart from the sign the coefficients in the first term of (2.9) and of the second term in (2.10) are the same. Equation (2.9) gives the polarization as the sum of two contributions, the piezoelectric due to the strain and the dielectric due to the field. Equation (2.10) states that the externally applied stress consists of two parts: first, that which would produce the prescribed strain if $\boldsymbol{E} = 0$, and second, that which is necessary to hold the strain constant when E is applied. From (2.9) we see that $e_{\alpha\beta\gamma}$ is the polarization per unit strain holding the macroscopic field $\boldsymbol{E}$ constant, and $\chi^c_{\alpha\beta}$ is the "clamped" dielectric susceptibility, i.e. for constant deformation ε (corresponding to a crystal under constant volume). In a non-piezoelectric crystal the first term in (2.9) and the second term in (2.10) vanish and (2.10) reduces to the purely elastic relation relating stress and strain where the coefficients $C_{\alpha\beta,\gamma\lambda}$ are the elastic constants as discussed in [Ref. 2.6, Sect. 3.6].

Substituting (2.10) into (2.7) and comparing the result with (2.9) we obtain relations between the piezoelectric constants e and the piezoelectric moduli d, as well as relations between the free and clamped susceptibilities χ^f and χ^c. Straightforward calculation gives

$$e_{\alpha\lambda\mu} = \sum_{\beta\gamma} d_{\alpha\beta\gamma} C_{\beta\gamma,\lambda\mu} \ , \tag{2.11}$$

$$\chi^f_{\alpha\lambda} = \chi^c_{\alpha\lambda} + \sum_{\beta\gamma} e_{\lambda\beta\gamma} d_{\alpha\beta\gamma} \ . \tag{2.12}$$

Similarly, substitution of (2.8) into (2.9) and comparison with (2.7) gives

$$d_{\alpha\lambda\mu} = \sum_{\beta\gamma} e_{\alpha\beta\gamma} S_{\beta\gamma,\lambda\mu} \ . \tag{2.13}$$

2.3 The Method of Homogeneous Deformations

We consider a crystal in the form of a slab perpendicular to the z-axis of the (x, y, z)-system. The unit cell contains two ions and each type of ion occupies planes parallel to the slab. The crystal therefore consists of a system of alternating planes occupied by cations and anions (Fig. 2.5).

The distances between neighbouring planes are not equal $(r>s)$ and the lattice constant is $a = r + s$. Such an arrangement of ions is found in crystals with cubic ZnS structure, where the planes containing Zn and S ions are perpendicular to the (111) direction of the cube (the z-direction in Fig. 2.5). Neighbouring planes are coupled by effective force constants f and g [which include in addition to the repulsive forces also the short-range part of the Coulomb forces (Sect. 2.4)].

For a homogeneous longitudinal distortion with displacements $u\binom{l}{\kappa}$ parallel to the z-axis (l: number of unit cell, $\kappa = 1, 2$: numbers the atom in a unit cell), the potential energy is

$$\Phi = \tfrac{1}{2} f \sum_l \left[u\binom{l}{2} - u\binom{l}{1} \right]^2 + \tfrac{1}{2} g \sum_l \left[u\binom{l+1}{1} - u\binom{l}{2} \right]^2 \tag{2.14}$$

for nearest neighbour interactions [Ref. 2.6, p. 17]. The elastic energy of unit cell l,

$$\Phi_{uc} = \tfrac{1}{2} f \left[u\binom{l}{2} - u\binom{l}{1} \right]^2 + \tfrac{1}{2} g \left[u\binom{l+1}{1} - u\binom{l}{2} \right]^2 \ , \tag{2.15}$$

is independent on l due to translational symmetry, and $\Phi = N\Phi_{uc}$, N being the number of unit cells ($N \to \infty$). Introducing the energy density $w = \Phi_{uc}/v_a$, where v_a is the volume of the unit cell we have

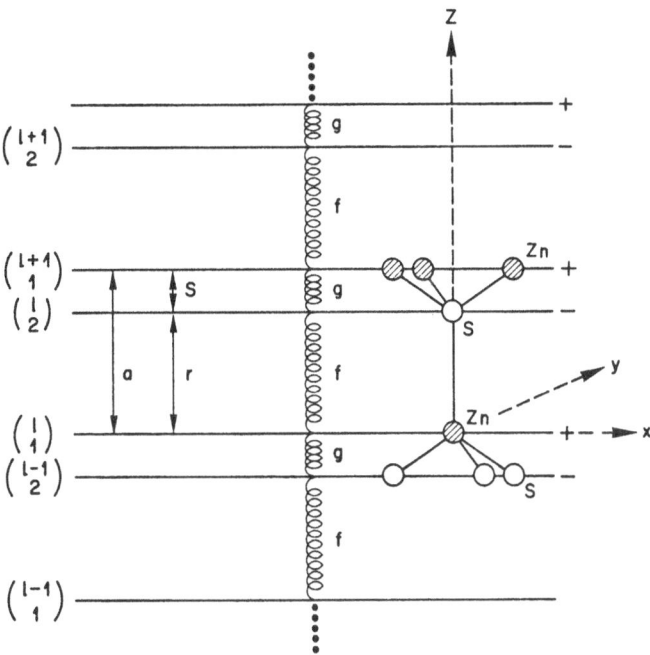

Fig. 2.5. One-dimensional model used to illustrate piezoelectricity. The planes are occupied by positive (+) or negative (−) ions and the distances between neighbouring planes are different ($r > s$). The inset shows the relation to the zinc-blende structure

$$w = \frac{1}{2v_a} \left\{ f \left[u\left(\tbinom{l}{2}\right) - u\left(\tbinom{l}{1}\right) \right]^2 + g \left[u\left(\tbinom{l+1}{1}\right) - u\left(\tbinom{l}{2}\right) \right]^2 \right\} . \tag{2.16}$$

We now assume that the energy density w corresponds to a crystal which is in a state of *homogeneous deformation*. This is a deformation in which each unit cell is altered in the same way. We can also say that if a lattice is subjected to a homogeneous deformation the resulting structure is still a perfect periodic structure; such homogeneous deformations are usually produced if a crystal is subjected to a stress. A homogeneous deformation can be built up as follows: First we subject the coordinates of all particles in the lattice to a linear homogeneous transformation: a particle initially at $\mathbf{r}$ is moved to $\mathbf{r}'$ where

$$r'_\alpha = r_\alpha + \sum_\beta \varepsilon_{\alpha\beta} r_\beta . \tag{2.17}$$

Here $\alpha, \beta = x, y, z$ and $\varepsilon_{\alpha\beta}$ are the components of the deformation tensor introduced in Sect. 2.2. If there are n atoms per unit cell ($\kappa = 1, 2 \ldots n$), we may further displace all particles of each sublattice κ by the same vector $\mathbf{u}_\kappa$

without destroying the periodicity of the lattice, although the structure will change in general. A particle $\left(\begin{smallmatrix} l \\ \kappa \end{smallmatrix}\right)$ is thus subjected to the total displacement

$$u_\alpha \left(\begin{smallmatrix} l \\ \kappa \end{smallmatrix}\right) = u_{\kappa\alpha} + \sum_\beta \varepsilon_{\alpha\beta} r_\beta \left(\begin{smallmatrix} l \\ \kappa \end{smallmatrix}\right) \; , \quad \text{where} \tag{2.18}$$

$$r_\beta \left(\begin{smallmatrix} l \\ \kappa \end{smallmatrix}\right) = r_\beta(l) + r_\beta(\kappa) \; .$$

For a Bravais lattice $u_{\kappa\alpha} = 0$ and (2.18) corresponds to Eq. (3.113) in Ref. [2.6]. Confining ourselves to the one-dimensional model with displacements in the z-direction and dropping the index z we can write

$$u \left(\begin{smallmatrix} l \\ \kappa \end{smallmatrix}\right) = u_\kappa + \varepsilon r \left(\begin{smallmatrix} l \\ \kappa \end{smallmatrix}\right) \; , \tag{2.19}$$

where $\varepsilon = (\partial u / \partial z)_0$ is the deformation in the z-direction. If the origin of the coordinate system is located at $\left(\begin{smallmatrix} 0 \\ 1 \end{smallmatrix}\right)$ the displacements for the two particles per unit cell are given by

$$u \left(\begin{smallmatrix} l \\ 1 \end{smallmatrix}\right) = u_1 + \varepsilon r(l) \; ,$$

$$u \left(\begin{smallmatrix} l \\ 2 \end{smallmatrix}\right) = u_2 + \varepsilon [r(l) + r] \; , \tag{2.20}$$

where $r(l) = la$, $r(1) = 0$, and $r(2) = r$ (Fig. 2.5). Substitution of (2.20) into (2.16) yields the energy density as a function of the deformation ε and the inner displacements u_1 and u_2, namely

$$w = \frac{1}{2v_a} [f(u_1 - u_2 - r\varepsilon)^2 + g(u_1 - u_2 + s\varepsilon)^2] \; , \tag{2.21}$$

where $s = a - r$ (Fig. 2.5). This expression is a generalization of the expression (3.123) in Ref. [2.6], which is valid for a Bravais lattice or for lattices in which each ion is at a centre of inversion for which the relative sublattice displacement $\delta = u_1 - u_2$ vanishes.

Corresponding to the two kinds of deformation, the inner displacements u_κ and the deformation ε, individual forces F_κ and a stress σ appear during the deformation; as derivatives of the potential energy, these quantities are given by[3]

$$F_\kappa = \frac{dw}{du_\kappa} \; , \quad \sigma = \frac{dw}{d\varepsilon} \; . \tag{2.22}$$

---

[3] Since we have here forces which must be exerted in deformations against the restoring forces in the crystal, they must be defined as positive derivatives of the energy density.

From (2.21, 22) we obtain

$$F_1 = \frac{1}{v_a}[(f+g)\delta - (rf - sg)\varepsilon] = -F_2 , \qquad (2.23a)$$

$$\sigma = -\frac{1}{v_a}[(rf - sg)\delta - (r^2 f + s^2 g)\varepsilon] , \qquad (2.23b)$$

where we have introduced the relative displacement of the two sublattices

$$\delta = u_1 - u_2 . \qquad (2.24)$$

From (2.23a) it follows that $F_1 + F_2 = 0$. External forces will therefore produce a homogeneous deformation only if the resultant force on a unit cell vanishes. This is the case if the forces F_κ originate from homogeneous electric fields. From (2.23a) we obtain

$$\delta = \frac{1}{f+g}[(rf - sg)\varepsilon + v_a F_1] . \qquad (2.25)$$

This equation shows that even in the absence of external forces $(F_\kappa = 0)$ a deformation ε will in general produce a relative displacement δ of the two sublattices if $rf \neq sg$. Substituting (2.25) into (2.23b) gives

$$\sigma = \frac{a^2}{v_a}\frac{fg}{f+g}\varepsilon - \frac{rf - sg}{f+g}F_1 . \qquad (2.26)$$

Identifying F_1 with the force per unit volume acting on the cation with effective charge e^* by a macroscopic field E in the z-direction, $F_1 = e^* E/v_a$, we obtain

$$\sigma = \frac{a^2}{v_a}\frac{fg}{f+g}\varepsilon - \frac{e^*}{v_a}\frac{rf - sg}{f+g}E . \qquad (2.27)$$

For the polarization $P = e^*\delta/v_a$ we obtain

$$P = \frac{e^*}{v_a}\frac{rf - sg}{f+g}\varepsilon + \frac{e^{*2}}{v_a(f+g)}E . \qquad (2.28)$$

These equations are compared with the macroscopic equations (2.9, 10) which in the one-dimensional case reduce to scalar equations and have the form

$$\sigma = C\varepsilon - eE , \qquad (2.29)$$

$$P = e\varepsilon + \chi^c E . \qquad (2.30)$$

15

Comparing (2.27, 28) with (2.29, 30) we obtain the following microscopic expressions for the elastic stiffness constant C the piezoelectric constant e and the static "clamped" susceptibility χ^c:

$$C = \frac{a^2}{v_a} \frac{fg}{f + g} \quad , \tag{2.31}$$

$$e = \frac{e^*}{v_a} \frac{rf - sg}{f + g} \quad , \tag{2.32}$$

$$\chi^c = \frac{e^{*2}}{v_a(f + g)} \quad . \tag{2.33}$$

Writing (2.7, 8) in the form

$$P = d\sigma + \chi^f E \quad , \tag{2.34}$$

$$\varepsilon = S\sigma + dE \quad , \tag{2.35}$$

the coefficients S, d and χ^f are obtained by comparison of (2.34, 35) with (2.29, 30) or directly from (2.11–13); using in addition (2.31–33), one obtains

$$S = C^{-1} = \frac{v_a}{a^2} \frac{f + g}{fg} \quad , \tag{2.36}$$

$$d = \frac{e^*}{a^2} \frac{rf - sg}{fg} \quad , \tag{2.37}$$

$$\chi^f = \chi^c \left[1 + \frac{(rf - sg)^2}{a^2 fg} \right] \quad . \tag{2.38}$$

A discussion of these results will be given in Sects. 2.4–6.

2.4 The Method of Long Waves

It is possible to establish a direct relation between piezoelectricity and lattice dynamics by using the method of long waves. In this approach the starting point consists in setting up the equations of motion and to consider the limiting case of the long-wavelength acoustic lattice waves with frequencies $\omega_j(\boldsymbol{q})$ as $\boldsymbol{q} \to 0$. These long wave acoustic modes represent elastic waves in a medium which is in a state of homogeneous deformations.

Rather than developing the general theory which requires lengthy calculations, we confine ourselves again to the one-dimensional model studied in Sect. 2.3 which contains all important physical aspects. For this case the macroscopic electromechanical equations are given by (2.29, 30). In particular, we have to derive an expression for the polarization from lattice dynamics, and compare it with the macroscopic result

$$P_0 = iqeu_0 + \chi^c E_0 \; , \tag{2.39}$$

which follows from (2.30) on substituting $\varepsilon = \partial u / \partial z$,

$$u = u_0 e^{i(qz - \omega t)} \tag{2.40}$$

and similar expressions for P and E.

In the microscopic approach the equations of motion are

$$m_\kappa \ddot{u} \binom{l}{\kappa} = - \sum_{l'\kappa'} \tilde{\Phi} \binom{ll'}{\kappa\kappa'} u \binom{l'}{\kappa'} + e_\kappa^* E^{(C)} \binom{l}{\kappa} \; . \tag{2.41}$$

Here $\tilde{\Phi} \binom{ll'}{\kappa\kappa'}$ are the usual (repulsive) short-range force constants, e_1^* is an effective charge of the ion κ $(e_1^* = e^*, e_2^* = -e^*)$, and $E^{(C)} \binom{l}{\kappa}$ is the total Coulomb field acting on the ion $\binom{l}{\kappa}$. This field has been discussed in detail in [Ref. 2.6, Sect. 4.2.2]; there we have seen that $E^{(C)}$ can be split up into the Lorentz field $E^{(L)}$, the "self-field" $E^{(S)}$ and the macroscopic field $E^{(M)}$. The expressions (4.54, 55) in [Ref. 2.6] for $E^{(L)}$ and $E^{(S)}$ converge rapidly, and consequently the corresponding forces $e_\kappa^* E^{(L)}$ and $e_\kappa^* E^{(S)}$ are essentially short-range forces which can be absorbed into effective short range force constants $\Phi \binom{ll'}{\kappa\kappa'}$. This is not possible for the force $e_\kappa^* E^{(M)}$ associated with the macroscopic field $E^{(M)} = E \binom{l}{\kappa}$, and the equation of motion is therefore written in the form

$$m_\kappa \ddot{u} \binom{l}{\kappa} = - \sum_{l'\kappa'} \Phi \binom{ll'}{\kappa\kappa'} u \binom{l'}{\kappa'} + e_\kappa^* E \binom{l}{\kappa} \; . \tag{2.42}$$

At this point it should be remembered that the macroscopic field E in (2.42) depends itself on the atomic displacements, that is, it is established by the ionic motion [Ref. 2.6, Eq. (4.50)]. Despite this fact there are two reasons to separate the contribution of the macroscopic field explicitly in the treatment of the long-wavelengths acoustic modes. Firstly, E appears also in the macroscopic relations (2.29, 30 and 39) with which we want to make contact, and secondly, the macroscopic field, as we know, has no unique value as $q \to 0$ but depends on the shape of the sample; for a slab shaped sample in the capacitor arrangement, the macroscopic field in the sample is equal to the original external field without a dielectric between the plates [2.22]. It should be remembered, that the macroscopic field plays also a distinct role in the $q \cong 0$ optical modes, because it is this field which is responsible for the splitting of the TO and LO modes [Ref. 2.6, Sect. 4.2.3].

After substituting

$$u\binom{l}{\kappa} = u_\kappa(q) \exp\{i[qr\binom{l}{\kappa} - \omega(q)t]\} \quad \text{and}$$

$$E = E_0(q) \exp\{i[qr\binom{l}{\kappa} - \omega(q)t]\}$$

into (2.42) and relating the force constants $\Phi\binom{ll'}{\kappa\kappa'}$ with the effective force constants f and g introduced in Fig. 2.5 one obtains [2.23]

$$\omega^2(q) \begin{pmatrix} m_1 & 0 \\ 0 & m_2 \end{pmatrix} \begin{pmatrix} u_1(q) \\ u_2(q) \end{pmatrix}$$

$$= \begin{pmatrix} f+g & -(fe^{iqr}+ge^{-iqs}) \\ -(fe^{-iqr}+ge^{iqs}) & f+g \end{pmatrix} \begin{pmatrix} u_1(q) \\ u_2(q) \end{pmatrix}$$

$$+ e^* E_0(q) \begin{pmatrix} -1 \\ 1 \end{pmatrix} = 0 , \tag{2.43}$$

i.e. of the form

$$\omega^2 mu = \overline{C}u + e^* E . \tag{2.44}$$

For acoustic modes ω goes to zero as $q \to 0$ and the limit has to be approached with care. The standard perturbation technique may be used and accordingly we replace q by $\tilde{\lambda}q$ where $\tilde{\lambda}$ is an expansion parameter, expand the q-dependent quantitites in (2.44) in powers of $\tilde{\lambda}$ and finally set $\tilde{\lambda} = 1$. Thus we have

$$\omega = \tilde{\lambda}\omega^{(1)} + \tfrac{1}{2}\tilde{\lambda}^2\omega^{(2)} + \ldots , \tag{2.45}$$

$$u_\kappa = u_\kappa^{(0)} + i\tilde{\lambda}u_\kappa^{(1)} - \tfrac{1}{2}\tilde{\lambda}^2 u_\kappa^{(2)} + \ldots , \tag{2.46}$$

$$E_0 = E_0^{(0)} + i\tilde{\lambda}E_0^{(1)} - \tfrac{1}{2}\tilde{\lambda}^2 E_0^{(2)} + \ldots , \tag{2.47}$$

and

$$\overline{C} = (f+g) \begin{pmatrix} 1 & -1 \\ -1 & 1 \end{pmatrix} + i\tilde{\lambda}q(sg - rf) \begin{pmatrix} 0 & 1 \\ -1 & 0 \end{pmatrix} + \ldots . \tag{2.48}$$

Note that there is no term $\omega^{(0)}$ in (2.45) since $\omega \to 0$ as $q \to 0$. These expressions are substituted into (2.44) and successive orders of the perturbation separated in the usual way. The zero-order result is

$$(f+g) \begin{pmatrix} 1 & -1 \\ -1 & 1 \end{pmatrix} \begin{pmatrix} u_1^{(0)} \\ u_2^{(0)} \end{pmatrix} + e^* E_0^{(0)} \begin{pmatrix} -1 \\ 1 \end{pmatrix} = 0 ,$$

i.e.

$$(f + g) \left(u_1^{(0)} - u_2^{(0)} \right) = e^* E_0^{(0)} \ . \tag{2.49}$$

Equating terms linear in $\tilde{\lambda}$ we obtain the first-order result

$$q(sg - rf) \begin{pmatrix} 0 & 1 \\ -1 & 0 \end{pmatrix} \begin{pmatrix} u_1^{(0)} \\ u_2^{(0)} \end{pmatrix} + (f + g) \begin{pmatrix} 1 & -1 \\ -1 & 1 \end{pmatrix} \begin{pmatrix} u_1^{(1)} \\ u_2^{(1)} \end{pmatrix}$$

$$+ \ e^* E_0^{(1)} \begin{pmatrix} -1 \\ 1 \end{pmatrix} = 0 \ . \tag{2.50}$$

Addition of the two equations in (2.50) shows that $u_1^{(0)} - u_2^{(0)} = 0$ (and hence $E_0^{(0)} = 0$). This result reflects the fact, that for vanishing q there is no relative displacement of the two atoms in the unit cell. In first order we have from (2.50) writing $u_1^{(0)} = u_2^{(0)} = u$,

$$u_1^{(1)} - u_2^{(1)} = \frac{rf - sg}{f + g} qu + \frac{e^*}{f + g} E_0^{(1)} \ . \tag{2.51}$$

To make contact with (2.39) we consider a macroscopic region which is small compared to the wavelength λ, and within which the polarization, from (2.46, 51), is given by

$$P_0 v_a = \sum_\kappa e_\kappa^* u_\kappa = \sum_\kappa e_\kappa^* u_\kappa^{(0)} + i\tilde{\lambda} \sum_\kappa e_\kappa^* u_\kappa^{(1)}$$

$$= u \sum_\kappa e_\kappa^* + i\tilde{\lambda} e^* qu \frac{rf - sg}{f + g} + i\tilde{\lambda} \frac{e^{*2}}{f + g} E_0^{(1)} \ .$$

The first term is zero due to the neutrality of the unit cell, so that, setting $\tilde{\lambda} = 1$, we obtain

$$P_0 = \frac{iqe^* u}{v_a} \frac{rf - sg}{f + g} + \frac{e^{*2}}{v_a(f + g)} i E_0^{(1)} \ . \tag{2.52}$$

From this equation we see that since $q = 2\pi/\lambda$, the piezoelectric contribution to the polarization is proportional to ξ/λ where ξ is a length of the order of a lattice parameter. For $\lambda \to \infty$ (pure translation of the crystal) this contribution disappears and only the term from the static susceptibility remains.

Comparing (2.52) with (2.39) we see that u and $iE_0^{(1)}$ in (2.52) have to be identified with u_0 and E_0, respectively, in (2.39), and that the piezoelectric constant e and the "clamped" static dielectric susceptibility χ^c are given by

$$e = \frac{e^*}{v_a} \frac{rf - sg}{f + g} \quad , \tag{2.53}$$

$$\chi^c = \frac{e^{*2}}{v_a(f + g)} \quad . \tag{2.54}$$

These expressions are identical with the expressions (2.32, 33) obtained by the method of homogeneous deformation (static approach).

Equation (2.29) can also be brought into a form suitable for comparison with the one-dimensional microscopic theory. It is merely necessary to differentiate with respect to z and write $\varrho\ddot{u} = \partial\sigma/\partial z$ where $\varrho = (m_1 + m_2)/v_a$ is the density. The coefficients C and e in (2.29) may then be derived by taking the microscopic theory to second order in $\tilde{\lambda}$ (Problem 2.1). The same results are obtained as those given by (2.31 and 32), respectively. The results (2.32, 37, 53) indicate that the piezoelectric constant e and the piezoelectric modulus d are zero if either the effective charge e^* vanishes or if $rf = sg$. The latter condition is usually satisfied if the distances between the planes are the same ($r = s$) such as in the NaCl structure [planes perpendicular to the (111) direction] or in the CsCl structure [planes perpendicular to the (100) direction]; in these cases every ion is located at a centre of inversion and no piezoelectricity will be observed. In the diamond structure which is obtained from the ZnS structure by replacing Zn and S bei C, $rf \neq sg$ but $e^* = 0$. Remember that in the diamond structure there is a centre of inversion midway between two nearest carbon atoms. On the other hand the effective charge will certainly be different from zero in ionic crystals, and for this reason piezoelectricity is most often observed in ionic or partly ionic crystals such as ZnS or ZnO. It should, however, be emphasized that there are at least three mechanisms which lead to $e^* \neq 0$ and contribute to the piezoelectric effect [2.24]:

1.   The displacement of the ionic charges.
2.   The internal displacements of the electron relative to the ion cores (electronic polarizabilities).
3.   The effect due to strain-induced change of the ionicity of the chemical bond (charge transfer effects).

In our simple model discussed above we have essentially considered only the displacements of the ionic charges and allowed for the other mechanisms only indirectly by introducing an effective charge e^* which may be different from the nominal ionic charge. In Sect. 2.1 we have already mentioned the extreme cases of trigonal Se and Te, in which the strong piezoelectric effect is probably due to the motion of bond charges or shells relative to the "ion cores". The microscopic origin of piezoelectricity is still incompletely known in many cases and effective piezoelectric charges are calculated on the basis of more modern theories, such as pseudopotential calculations, bond-orbital models and bond charge models [2.25–29, 38].

2.5 Application to Crystals with ZnS-Structure

It is possible to relate the effective interplane coupling force constants f and g introduced in the one-dimensional model (Fig. 2.5) with effective valence force constants of the ZnS structure which are known from infrared absorption and Raman scattering experiments. These force constants are introduced in Fig. 2.6; the relation of the basic unit of ZnS with the one-dimensional model is also illustrated in Fig. 2.5. K is the Zn-S stretching force constant, H_1 the S-Zn-S and H_2 the Zn-S-Zn bending force constant [Ref. 2.6, Sect. 4.5]. In order to construct the dynamical matrix the internal coordinates (changes in bond lengths and bond angles) are transformed to cartesian coordinates $u\binom{l}{\kappa}$ by using the procedure described in [Ref. 2.6, Sect. 4.5]. For $q \| z$ a 6×6 matrix is obtained which factors into three 2×2 matrices with the following elements ($\kappa = 1$ for Zn, $\kappa = 2$ for S, $H = H_1 + H_2$):

$$C_{\alpha\alpha}(11) = \frac{1}{3}(4K + 16H_1 + 13H_2) + H_2 \cos \frac{4qr}{3} ,$$

$$C_{\alpha\alpha}(22) = \frac{1}{3}(4K + 13H_1 + 16H_2) + H_1 \cos \frac{4qr}{3} ,$$

$$C_{\alpha\alpha}(12) = -2He^{iqr} - \tfrac{2}{3}(2K - 5H)e^{-iqs} ,$$

$$C_{\alpha\alpha}(21) = C_{\alpha\alpha}^*(12) ,$$

$$(2.55)$$

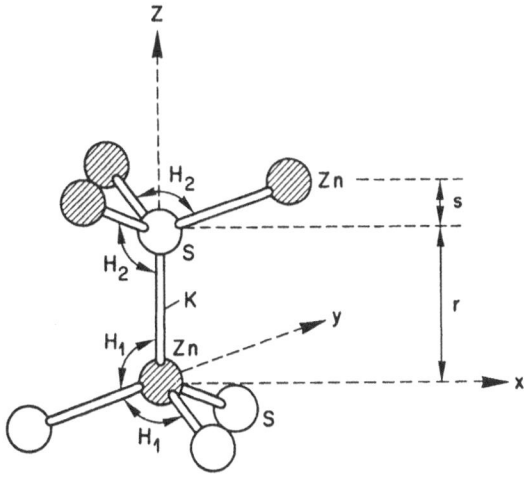

Fig. 2.6. Basic group of ions in the zinc-blende structure. K is the stretching force constant, H_1 and H_2 are the bending force constants

where $\alpha = x, y$. For $\alpha = z$ one finds

$$C_{zz}(11) = C_{zz}(22) = \tfrac{4}{3}(K + 4H) \ ,$$
$$C_{zz}(12) = -Ke^{iqr} - \tfrac{1}{3}(K + 16H)e^{-iqs} \ ,$$
$$C_{zz}(21) = C_{zz}^{*}(12) \ . \tag{2.56}$$

For $q = 0$ these expressions are identical with those derived by *Tsuboi* [2.30]. Comparing the matrix $C_{zz}(q)$ given by (2.56) with the corresponding matrix $\overline{C}$ in (2.43, 44), and noting that in the latter $s = r/3$ for the ZnS structure, we obtain

$$f_z = f = K \ , \quad g_z = g = \tfrac{1}{3}(K + 16H) \ . \tag{2.57}$$

Substituting these values into (2.53) we obtain

$$e = \frac{2}{\sqrt{3}} \frac{e^{*}}{a_0^2} \frac{K - 2H}{K + 4H} \ , \tag{2.58}$$

where we have used $v_a = a_0^3/4$, $r = \sqrt{3}a_0/4$, and a_0 is the lattice constant of ZnS.

Until now we have treated a one-dimensional problem for which the piezoelectric tensor reduces to a scalar. The piezoelectric constant e corresponds to a strain deformation ε_{zz} along the z-axis which is parallel to the (111) direction of the cube, and we write $e = e'_{333} = e'_{zzz}$[4]. It is now easy to evaluate also the piezoelectric constants $e'_{113} = e'_{xxz}$ and $e'_{223} = e'_{yyz}$ associated with the shearing deformations ε_{xz} and ε_{yz}, respectively, for the case $q \| z$. These piezoelectric constants can be obtained from equations of motion similar to (2.43, 44), namely

$$\omega^2 m u_\alpha = \overline{C}_{\alpha\alpha} u_\alpha + e^{*} E_\alpha \tag{2.59}$$

where $\alpha = x$ or y. The matrix $\overline{C}_{\alpha\alpha}$ has the same structure as the matrix $\overline{C}$ appearing in (2.43) but with different force constants: $f = f_z$ is replaced by $f_x = f_y$ and $g = g_z$ by $g_x = g_y$. In (2.59) the macroscopic field E_α is induced by the shear wave $\varepsilon_{\alpha z}$ ($\alpha = x, y$) which causes compressions (or dilatations) along the other three cube diagonals of the ZnS structure. The matrix $\overline{C}_{\alpha\alpha}$ in (2.59) cannot be directly compared with the matrix $C_{\alpha\alpha}$ defined by (2.55). In the latter, terms of the form $H_1\cos(4qr/3)$ and $H_2\cos(4qr/3)$ appear which describe the coupling between nearest Zn-Zn

---

[4] In the following the components e'_{ijk} refer to the coordinate system (x, y, z) with $z \| (111)$ (Fig. 2.6), while the components e_{lmn} refer to the coordinate system (x_1, x_2, x_3) with $x_1 \| (100)$, $x_2 \| (010)$ and $x_3 \| (001)$.

and S-S planes separated by a distance $4r/3$; in $\overline{C}_{\alpha\alpha}$ no such interactions are introduced. A direct comparison between $C_{\alpha\alpha}$ and $\overline{C}_{\alpha\alpha}$ is only possible for $qr \ll 1$ as relevant for piezoelectricity; in this case $\cos(4qr/3) \cong 1$ and the comparison yields

$$f_x = f_y = 2H \ , \quad g_x = g_y = \tfrac{2}{3}(2K + 5H) \ . \tag{2.60}$$

Substituting these values into (2.53) we obtain

$$e'_{113} = e'_{223} = -\frac{1}{2}e'_{333} = -\frac{1}{\sqrt{3}}\frac{e^*}{a_0^2}\frac{K - 2H}{K + 4H} \ . \tag{2.61}$$

The piezoelectric constants of ZnS are usually not referred to the system (x, y, z) with $z \| (111)$ but rather to the system (x_1, x_2, x_3) with axis parallel to the edges of the cube. The transformation between the two coordinate systems is given by

$$\begin{pmatrix} x \\ y \\ z \end{pmatrix} = \begin{pmatrix} a_{11} & a_{12} & a_{13} \\ a_{21} & a_{22} & a_{23} \\ a_{31} & a_{32} & a_{33} \end{pmatrix} \begin{pmatrix} x_1 \\ x_2 \\ x_3 \end{pmatrix}$$

$$= \begin{pmatrix} -\frac{1}{\sqrt{6}} & -\frac{1}{\sqrt{6}} & \sqrt{\frac{2}{3}} \\ \frac{1}{\sqrt{2}} & -\frac{1}{\sqrt{2}} & 0 \\ \frac{1}{\sqrt{3}} & \frac{1}{\sqrt{3}} & \frac{1}{\sqrt{3}} \end{pmatrix} \begin{pmatrix} x_1 \\ x_2 \\ x_3 \end{pmatrix} \ , \tag{2.62}$$

and the tensor components in the two systems are related as follows [2.3]

$$e'_{ijk} = \sum_{lmn} a_{il}a_{jm}a_{kn}e_{lmn} \ . \tag{2.63}$$

For the ZnS structure only the components $e_{123} = e_{132} = e_{231} = e_{213} = e_{312} = e_{321}$ are different from zero [2.3] (in Voigt's notation this is the component e_{14}). From (2.62, 63) one finds

$$e = e'_{333} = \frac{2}{\sqrt{3}}e_{123} \quad \text{and} \tag{2.64}$$

$$e'_{113} = e'_{223} = -\frac{1}{\sqrt{3}}e_{123} \ . \tag{2.65}$$

In addition, one obtains

$$e'_{111} = \sqrt{\frac{2}{3}}e_{123} \ , \tag{2.66}$$

$$e'_{222} = 0 \ . \tag{2.67}$$

23

The latter two constants are associated with strains parallel to the x and y axis and correspond to acoustic waves with q parallel to the x and y axis, respectively. From (2.67) we note that a compression along the y axis which is parallel to a face diagonal of the cube, does not give rise to piezoelectricity because the associated compressions and dilatations along the four space diagonals cancel. From (2.58) we can make a rough estimate of the effective piezoelectric charge e^*. With $a_0 = 5.4093\,\text{Å}$, $K \cong 1.267\,\text{mdyn/Å}$, $H \cong 0.1$ mdyn/Å [2.36], and $e = (2/\sqrt{3})\,e_{123} = (2/\sqrt{3})\,e_{14} = 0.168\,\text{C/m}^2$, we obtain $e^* \cong 0.42\,e_0$, where e_0 is the charge of the electron. Thus, our simple lattice dynamical model which is based on the valence force field predicts a reasonable magnitude for the effective charge. At this point it should be mentioned that the piezoelectric effective charge is in general different from the optical effective charge (the Szigeti charge, for example [Ref. 2.6, Sect. 4.3.1]) because the effective charges depend on the mode in question. This point has been discussed on the basis of the bond-orbital model by *Harrison* [2.25]. In Problem 2.6.3 the effective piezoelectric charge and the effective optic charge are evaluated on the basis of a one-dimensional shell-model. For more detailed studies of the piezoelectric effect of crystals with ZnS structure the reader is referred to the literature [2.19, 28, 31–38].

2.6 Problems

2.6.1 The Method of Long Waves

Calculate the elastic constant C and the piezoelectric constant e in (2.29) by applying the method of long waves developed in Sect. 2.4. The same results are obtained as those given by (2.31, 32).

Hint: Differentiate (2.29) with respect to z and use $\varrho \ddot{u} = \partial\sigma/\partial z$, where $\varrho = (m_1 + m_2)/v_a$ is the density. Expand (2.43) to second order in $\tilde{\lambda}$ and use the zero-order result $u_1^{(0)} = u_2^{(0)} = u$ and the first-order result (2.51).

2.6.2 Elastic Constant of a Piezoelectric Crystal for Short-Circuit and Open-Circuit Conditions

Consider a flat plate of a piezoelectric crystal with the plane of the plate normal to the direction of σ, E and P considered in (2.29, 30). Find expressions for the effective elastic constants when the surfaces of the plate are short-circuited (C^E) and when they are electrically open (C^P).

Result:

$$C^E = \frac{a^2}{v_a}\frac{fg}{f+g} \qquad \text{according to (2.31).}$$

$$C^P = C^E + \frac{e^2}{\chi^c} = \frac{1}{v_a}(r^2 f + s^2 g) \ .$$

Note that $C^P > C^E$ due to the depolarization field which is present in open-circuit conditions.

Hint: From (2.29) we have $\sigma = C^E \varepsilon - eE$, where C^E is the elastic constant at constant field, for example at $E = 0$. C^E is given by (2.31). Substituting E from (2.30) into (2.29) gives $\sigma = \sigma(\varepsilon, P)$; C^P is the elastic constant at constant polarization P.

2.6.3 Piezoelectric and Optic Effective Charges for the Shell Model

Consider a crystal containing planes occupied by cations and anions [such as in ZnS, planes perpendicular to (111)] (Fig. 2.7). For simplicity only the anions are supposed to be polarizable. $u\binom{l}{\kappa}$, $v\binom{l}{\kappa}$ are the displacements of ion cores and shells, respectively, ze_0 is the charge of the cation; xe_0 and ye_0 are the charges of the core and shell of the anion, and $(x + y)e_0 = -ze_0$ is the charge of the anion. S, S' and T, T' are effective core-shell and core-core force constants, and k is the isotropic core-shell force constant of the anion. By using the method of long waves and the procedure described in [Ref. 2.6, Sect. 4.8.1], calculate the piezoelectric constant e and the effective charge $z^* e_0$ for the optical vibrations.

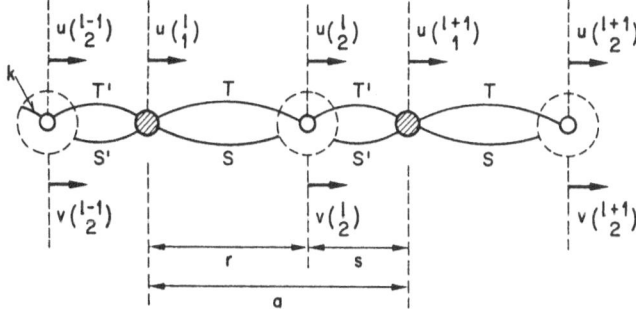

Fig. 2.7. One-dimensional piezoelectric chain with polarizable anions

Result:

$$e = \frac{rf - sf'}{f + f'} e^*_{\text{piez}} \ ,$$

$$e^*_{\text{piez}} = e_0 z_{\text{piez}} \ , \quad z_{\text{piez}} = z + a\alpha \frac{S'T - ST'}{rf - sf'} y \ ,$$

$$z^* = z + \alpha(S + S')y \ ,$$

25

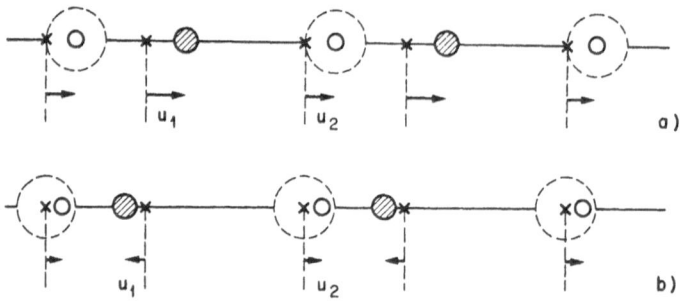

Fig. 2.8. (a) The piezoelectric mode is a long wavelength acoustic mode with small sublattice displacement $u_1 - u_2 \sim q$ and consequently small displacements between core and shells are expected. **(b)** In the $q \sim 0$ optic mode the sublattice displacement $u_1 - u_2$ is large which give rise to large displacements between core and shells

where

$$f = \alpha k S + T , \quad f' = \alpha k S' + T' , \quad \alpha = (S + S' + k)^{-1} .$$

Since $y < 0$, z^* is smaller than z. The second term in the expression for z_{piez} is usually small, and the shell model therefore predicts a piezoelectric charge which is close to the ionic charge $z e_0$. In fact, if $T = T' = 0$ one obtains $e^*_{piez} = z e_0$. This shows that e^*_{piez} and $e^*_{optic} = z^* e_0$ are in general quite different because the modes involved are different. In fact, for a long wavelength acoustic mode (Fig. 2.8a) with a small relative sublattice displacement $u_1 - u_2 \sim q$ we expect little displacements between shells and cores and the effective charge is essentially given by $z e_0$. In the $q = 0$ optic mode, however, large relative displacements between cores and shells are expected due to repulsion between the shells. It should be mentioned that the shell model is in general not suitable to describe piezoelectricity in that it overestimates the piezoelectric effect [2.36]; this is probably due to the fact that the shell model does not take into account charge redistribution (change in ionicity) [2.25, 38].

Hint:

1. Set up the equations of motion for cores and shells by using the macroscopic field instead of the effective field, as in (2.42).
2. Eliminate the shell coordinate.
3. Apply the method of long waves and calculate the polarization $P = v_a^{-1} (z u_1 + x u_2 + y v) e_0$; compare P with the macroscopic expression to obtain e.
4. Derive the equations of motion for the $q = 0$ optic modes by applying the procedure in [Ref. 2.6, Sect. 4.3.1, Eq. (4.82)] to obtain z^*.

3. Ferroelectricity

In this chapter we study the role of phonons in ferroelectric phase transitions. Section 3.1 is devoted to a description of general properties of ferroelectric materials. In Sect. 3.2 ferroelectric materials are classified on the basis of materials and structures, and the static dielectric properties of selected ferroelectrics are discussed. Section 3.3 contains the thermodynamic aspects of ferroelectric phase transitions on the basis of Devonshire's theory; first- and second-order transitions are discussed. In Sect. 3.4 the general aspects of the lattice dynamics of displacive ferroelectric phase transitions are presented, including the lattice dynamical basis of Devonshire's theory. Section 3.5 contains a discussion of some lattice dynamical models used for the study of ferroelectric phase transitions. The soft-mode spectroscopy of selected ferroelectrics is discussed in Sect. 3.6. Section 3.7 contains a short discussion of quantum ferroelectricity, while Sect. 3.8 is devoted to disordered polar systems. The last section contains a number of illustrative problems.

3.1 General Properties of Ferroelectric Materials

A ferroelectric crystal is a solid which, below a critical temperature, exhibits a spontaneous polarization, i.e., it is polarized in the absence of an external electric field; in general, the direction of the spontaneous polarization may be altered under the influence of an applied field. The direction of spontaneous polarization need not be the same throughout a macroscopic crystal. Rather, the crystal consists of a number of domains; within each domain the polarization has a specific direction, but this direction varies from one domain to another. Figure 3.1a illustrates the circuit for measuring the polarization as a function of the applied field, while Fig. 3.1b shows the results of such a measurement, namely a hysteresis loop which is typical for ferroelectric materials. On the basis of the domain concept, the occurrence of hysterisis in the P versus E relationship can be explained as follows: With reference to Fig. 3.1b, consider a crystal which initially has an overall polarization equal to zero, i.e., the sum of the vectors representing the dipole moments of the individual domains vanishes. When an electric field is applied to the crystal, the domains with polarization components along the applied field direction grow at the expense of the "antiparallel" domains; thus the polarization increases (OA). When all domains are aligned

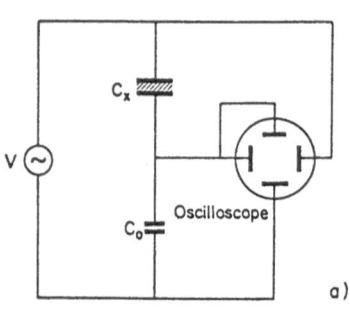
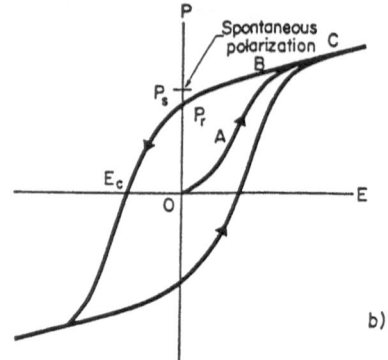

Fig. 3.1. (a) Circuit for display of ferroelectric hysteresis loop. The voltage across the ferroelectric crystal C_x is applied to the horizontal plates of the oscilloscope. The linear capacitor C_0 is in series with C_x. The voltage across C_0 is proportional to the polarization of C_x and is applied to the vertical plates. (b) Schematic representation of hysteresis loop in the polarization versus applied field relationship

in the direction of the applied field (BC), the polarization saturates and the crystal becomes single-domain. A further increase of the polarization with increasing applied field results from "normal" polarization effects as observed in normal dielectric crystals. The extrapolation of the linear part BC to zero external field gives the *spontaneous polarization* P_s. The value of P_s so obtained is evidently the same as the polarizations which already existed within each of the domains in the virgin state corresponding to the origin O in Fig. 3.1b. Thus, when we speak of "spontaneous polarization" we have in mind the polarization within a single domain and not the over-all polarization of a crystal. When the applied field for a crystal corresponding to point B in Fig. 3.1b is reduced, the polarization of the crystal decreases, but for zero applied field there remains the *remanent polarization* P_r, where P_r refers to the crystal as a whole. In order to remove the remanent polarization, the polarization of approximately half the crystal must be reversed and this occurs only when a field in the opposite direction is applied. The field required to reduce the polarization to zero is called the *coercive field* E_c.

At this point a few remarks may be made about the crystal structure of ferroelectrics. A necessary, but not sufficient condition for a solid to be ferroelectric is the absence of a center of symmetry. In total there are 21 classes of crystals which lack a center of symmetry. Of these 21 classes, 20 are piezoelectric (Chap. 2). Ten out of the 20 piezoelectric classes exhibit pyroelectric effects. These *pyroelectric crystals* are spontaneously polarized. However, the polarization is usually masked by surface charges which collect on the surface from the atmosphere; when the temperature of such a crystal is altered, the polarization changes and this change can be observed, hence the name pyroelectricity.

Ferroelectricity usually disappears above a certain critical temperature T_c; this temperature is called the *transition temperature* or the ferroelectric *Curie temperature*. T_c is the temperature at which the spontaneous polarization vanishes. Above the transition temperature the crystal is said to be in a *paraelectric state*. The term paraelectric suggests an analogy with paramagnetism and implies a rapid decrease in the dielectric constant, as the temperature increases. Associated with the transition from the ferroelectric to the nonferroelectric phase are anomalies in other physical properties. Thus, in a *first-order transition* there will be a discontinuous change in the structure and the volume, associated with a latent heat. In a *second-order transition*, however, the structure changes continuously, but the specific heat will exhibit a discontinuity.

The static dielectric constant ε_0 of a ferroelectric is, of course, not a constant, but depends on the applied field E; this is a consequence of the nonlinear relationship between P and E (Fig. 3.1b). When one speaks of "the dielectric constant", one refers to the slope of the curve OA at the origin, i.e. ε_0 is measured for small applied fields $(E \rightarrow 0)$ so that no motion of domain boundaries occurs. The static dielectric constant so defined,

$$\varepsilon_0 = 1 + 4\pi \lim_{E \rightarrow 0} (P/E) \ ,$$

is very large in the vicinity of the transition temperature, of the order of 10^4 to 10^5 [1]. Above the transition temperature ε_0 obeys the *Curie-Weiss law*

$$\varepsilon_0 = \frac{C'}{T - T_0} + \varepsilon_\infty \ , \tag{3.1}$$

where C' is a constant, and T_0 is defined as the temperature at which the extrapolated curve of the inverse dielectric constant cuts the temperature axis (Fig. 3.11). For a first-order transition $T_0 < T_c$, but for a second-order transition $T_0 = T_c$ (Sect. 3.3.2). ε_∞ is the high-frequency dielectric constant contributed by the electronic polarization [Ref. 3.1, Sect. 4.3.2]. In the vicinity of the transition temperature ε_∞ may be neglected. Likewise, the susceptibility $\chi = (\varepsilon - 1)/4\pi \cong \varepsilon/4\pi$ is given by

$$\chi = \frac{C}{T - T_0} \tag{3.2}$$

in this region, where $C = C'/4\pi$ is called the *Curie constant*. The principles and applications of ferroelectrics and polar dielectrics are summarized comprehensively by *Lines* and *Glass* [3.2a], and by *Burfoot* and *Taylor* [3.2b].

---

[1] A physical meaningful dielectric constant is also measured in the paraelectric state as $T \rightarrow T_0$ or on single domain crystals.

3.2 Classification and Properties of Selected Ferroelectrics

Ferroelectric crystals may be classified into several groups [3.3]. First there is Rochelle salt, $NaK(C_4H_4O_6)\cdot 4H_2O$, and related isomorphous salts. Rochelle salt is a complicated crystal; for a summary of properties see *Mueller* and *Mason* [3.4].

The second group of ferroelectric crystals is the KDP group, where KDP stands for KD_2PO_4. From an analysis of the structure of KH_2PO_4 it appears that the PO_4 groups form tetrahedrons with the four oxygens at the corners and the phosphorus at the center. These phosphate groups are bounded together by *hydrogen bonds*. From the large effect deuterion has on T_c it is clear that the occurrence of ferroelectricity is associated in large degree with proton tunneling and hydrogen bonding in these lattices. In a simple picture the protons in each hydrogen bond O-H...O move in a double-well potential whose minima are separated by 0.4 Å. Above T_c the hydrogen atoms are statistically distributed over the two sites (disordered structure). Below T_c the hydrogen atoms order preferentially on one of those sites (ordered structure). It should be emphasized, however, that the motion of the protons is not directly associated with the spontaneous polarization. This follows from the fact that the proton displacements at T_0 in KDP are nearly perpendicular to the direction of polarization, and it is therefore clear that some complicated coupling between the protons and the heavy ions must exist. At T_0 this coupling changes in such a way as to lead to a permanent displacement of the heavy ions below T_0 and hence to a spontaneous polarization. Figure 3.2 shows the spontaneous polarization P_s and the dielectric constant ε_0 of KH_2PO_4 [3.5,6]. KH_2PO_4 is tetragonal for $T>T_c$ and orthorhombic for $T<T_c$; the ferroelectric axis is the c-axis. Another group of ferroelectric crystals containing hydrogen

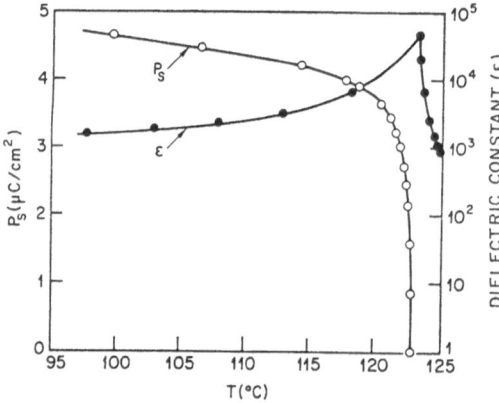

Fig. 3.2. Temperature dependence of the spontaneous polarization P_s and of the dielectric constant ε of KH_2PO_4. The dielectric constant is measured along the ferroelectric axis [3.5,6]

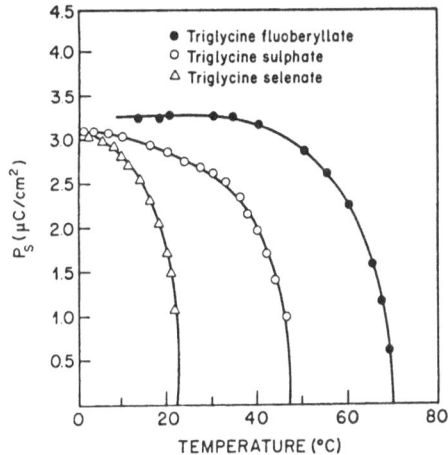

Fig. 3.3. Temperature dependence of the spontaneous polarization P_s of triglycine sulphate, triglycine fluoberyllate and triglycine selenate [3.7]

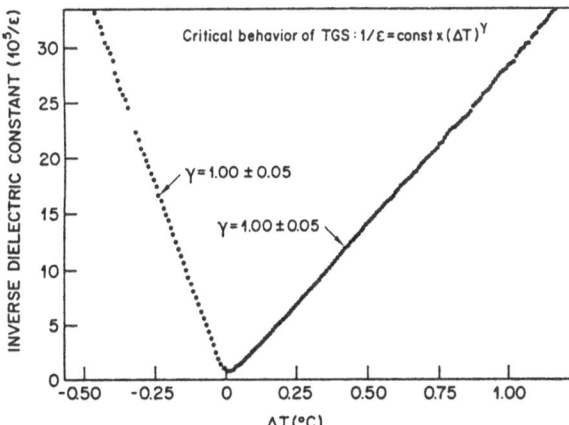

Fig. 3.4. Plot of the inverse dielectric constant of TGS versus temperature ($T_c = 49.42 \pm 0.05°C$) [3.8]

bonds and exhibiting an order-disorder transition is the TGS group, where TGS stands for tri-glycine sulfate or tri-glycine selenate. TGS is used as a pyroelectric detector. Figure 3.3 and 3.4 show the spontaneous polarization P_s and the inverse dielectric constant versus temperature [3.7,8]. The critical behaviour of TGS can be accounted for by [3.8]

$$P_s = \text{const}(\Delta T)^\beta \quad \text{and} \tag{3.3}$$

$$1/\varepsilon = \text{const}(\Delta T)^\gamma \ , \tag{3.4}$$

where $\Delta T = |T_c - T|$, $\beta = 0.51 \pm 0.05$ and $\gamma = 1.00 \pm 0.01$. For a second-order ferroelectric phase transition the thermodynamic theory predicts $\beta = 1/2$ and $\gamma = 1$ (Sect. 3.3.3).

31

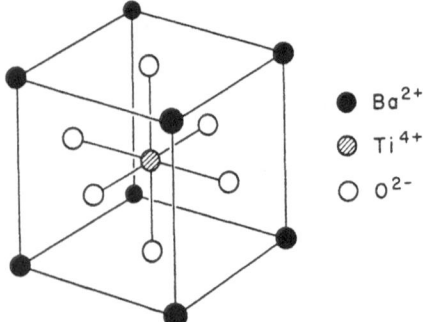

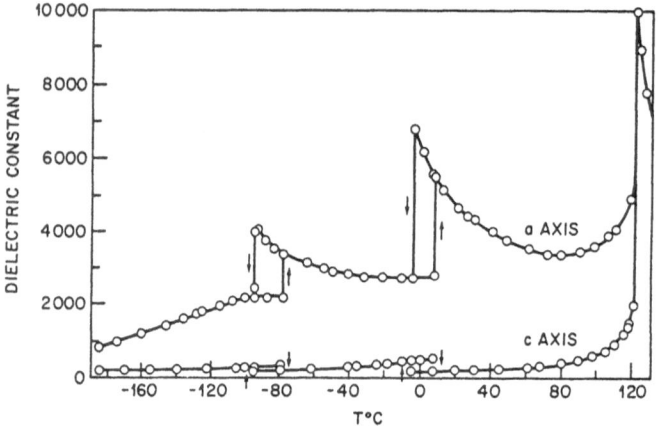

Fig. 3.6. Temperature dependence of the dielectric constant of BaTiO₃, measured along the pseudo-cubic [010] and [001] axis [3.9, 10]

The third group of ferroelectrics consists of ionic crystals with crystal structures closely related to the perovskite structure. Examples are $BaTiO_3$, $SrTiO_3$, $KNbO_3$ and $PbTiO_3$.

The perovskite structure is one of the simpler structures to exhibit ferroelectricity. Figure 3.5 shows the structure of $BaTiO_3$. In the unpolarized state the crystal is cubic with Ba^{2+} ions at the cube corners, O^{2-} ions at the centers of the cube faces, and Ti^{4+} ions at the cube centers. Below the Curie temperature of 393 K the structure is first deformed to a tetragonal structure, the positive ions being displaced relative to the negative ones along a [100] direction, thereby developing a polarization in this direction. Figure 3.6 shows the temperature dependence of the dielectric constant of $BaTiO_3$ [3.9, 10]. The discontinuities near 0°C and −80°C are caused by small changes in the crystal structure. The spontaneous polarization is parallel to a cube edge above 0°C, becomes parallel to a face diagonal below

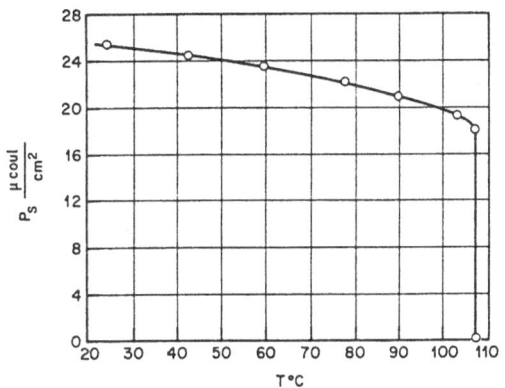

Fig. 3.7. Spontaneous polarization of $BaTiO_3$ in the tetragonal phase [3.11–13]

0°C, and parallel to a body diagonal below –80°C. The dielectric constant is finite at the transition point at 120°C and drops discontinuously. Figure 3.7 shows the temperature dependence of the spontaneous polarization of $BaTiO_3$ in the tetragonal phase [3.11–13]. Note the discontinuous increase of P_s at the transition point, which is followed by a continuous increase as the temperature is further lowered.[2] This discontinuity shows that the transition in $BaTiO_3$ is of first order, in contrast to the second-order transition of TGS (Fig. 3.3).

Finally it should be mentioned that narrow-gap semiconductors such as PbTe and SnTe and their alloys exhibit a tendency for ferroelectric displacive phase transitions from a high-temperature rocksalt structure to a rhombohedral phase at low temperatures [3.14]. In Fig. 3.8 results for the inverse quasi-static dielectric constant ε_s^{-1} are given as a function of temperature for two p-type samples of PbTe with different carrier concentrations. Here ε_s is designated as quasistatic, since measurements are performed in the MHz regime. At high temperatures, ε_s^{-1} varies linearly with temperature according to the Curie-Weiss law, whereas at low temperatures ε_s^{-1} saturates due to zero-point fluctuations. Figure 3.8 shows that PbTe does indeed exhibit a tendency for a ferroelectric phase transition but never really becomes ferroelectric. A similar situation is found for SnTe which is another example for a "near-ferroelectric"; the same is true for $SrTiO_3$ for which an extrapolated transition temperature $T_0 = 32\,\mathrm{K}$ is found (Fig. 3.21). Such "near-ferroelectrics" are also called *incipient ferroelectrics*. The situation is different for GeTe, which has the sodium-chloride structure above about 670 K and a trigonal structure below this temperature. These two structures are related by a displacive ferroelectric transition. In SnTe and GeTe the high conductivity prevents any direct measurement of the static dielectric constant.

---

[2] Because of domain effects, the early measurements often showed a continuous increase in P_s upon cooling below the 120°C transition point.

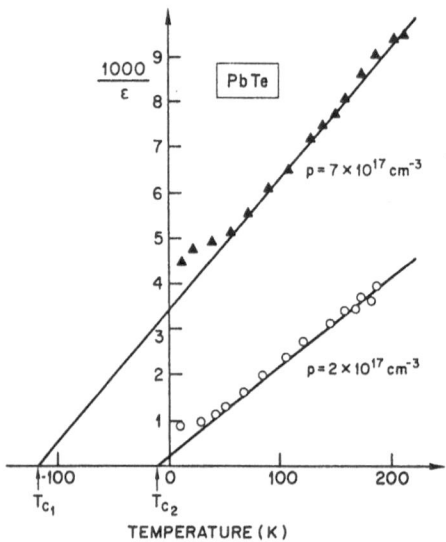

Fig. 3.8. Temperature dependence of the inverse quasi-static dielectric constant of two p-PbTe samples of different hole concentrations [3.14]

The above classification of ferroelectric crystals is based on a materials point of view. From a structural point of view ferroelectric crystals can also be grouped into one or the other of two categories – *displacive* or *order-disorder*. Crystals of the KDP and TGS groups are examples of order-disorder ferroelectrics, while the perovskites and the IV–VI semiconductors are examples of displacive ferroelectrics. Experience, however, tells us now that few, if any, ferroelectric crystals fall unequivocally into either of these classifications. Most real systems probably reside somewhere between the two extremes. Finally, from a thermodynamic point of view, ferroelectric transitions can be grouped into *first-order* and *second-order* transitions. $BaTiO_3$ is a crystal exhibiting a first-order transition (Fig. 3.7), while TGS and TSCC are crystals exhibiting a second-order transition (Figs. 3.4, 25).

3.3 Thermodynamic Theory of Ferroelectrics

3.3.1 General Considerations

The thermodynamic theory of ferroelectricity [3.15–18] does not interpret the macroscopic properties of ferroelectrics in terms of their microscopic properties. It only attempts to relate all the macroscopic properties to each other and to describe them in terms of a few phenomenological parameters. It is the purpose of Sects. 3.4, 5 to relate these parameters with the microscopic parameters of lattice dynamical models. Readers who are familiar with the thermodynamic theory of ferroelectric phase transitions (Landau-Devonshire theory) should skip this section and continue with Sect. 3.4.

34

The differential dU of the internal energy of a body subject to an electric field at zero stress is

$$dU = T\,dS + \boldsymbol{E}\cdot d\boldsymbol{P} ,\tag{3.5}$$

where S is the entropy, T the temperature, $\boldsymbol{E}$ the applied electric field and $\boldsymbol{P}$ the polarization. The Helmholtz free energy is $F = U - TS$, and from (3.5) we obtain

$$dF = -S\,dT + \boldsymbol{E}\cdot d\boldsymbol{P} .\tag{3.6}$$

In most dielectrics the polarization $\boldsymbol{P}$ is strictly proportional to the applied electric field $\boldsymbol{E}$. Dielectric breakdown occurs before saturation. In a ferroelectric crystal, however, the polarization can become so large that nonlinear effects become dominant. This means that even in a single-domain crystal[3] $\boldsymbol{P}$ is in general a nonlinear function of $\boldsymbol{E}$, and conversely, $\boldsymbol{E}$ is a nonlinear function of $\boldsymbol{P}$ [Ref. 3.12, p. 29]. The free energy can therefore be expanded in terms of the components of $\boldsymbol{P}$. If the crystal has a center of symmetry above the Curie temperature, only even terms are significant in this expansion. At constant temperature $dF = \boldsymbol{E}(\boldsymbol{P})\cdot d\boldsymbol{P}$, and we can write

$$F = F_0 + \tfrac{1}{2}a(P_x^2 + P_y^2 + P_z^2) + \tfrac{1}{4}b(P_x^4 + P_y^4 + P_z^4) \\ + \tfrac{1}{6}c(P_x^6 + P_y^6 + P_z^6) + \tfrac{1}{4}d(P_x^2 P_y^2 + P_x^2 P_z^2 + P_y^2 P_z^2) + \dots ,\tag{3.7}$$

where F_0 is the free energy for zero polarization, which is often equated to zero. Let us assume that in the ferroelectric phase the spontaneous polarization lies along the z-direction, and that electric fields are only applied along this direction, so that $P_x = P_y = 0$ and $P_z = P$. Equation (3.7) can then be rewritten as follows[4]

$$F = F_0 + \tfrac{1}{2}aP^2 + \tfrac{1}{4}bP^4 + \tfrac{1}{6}cP^6 ,\tag{3.8}$$

where the coefficients a, b and c in general depend on temperature. Differentiation of (3.8) with respect to P and using (3.6) gives the following equation for the electric field acting on a ferroelectric, in terms of the polarization P:

$$\partial F/\partial P = E = aP + bP^3 + cP^5 .\tag{3.9}$$

For $E = 0$ the conditions for stability are

$$\frac{dF}{dP} = 0 , \quad \frac{d^2 F}{dP^2} \geq 0 .\tag{3.10}$$

---

[3] Nonlinear effects associated with the domain structure are excluded here.

[4] Equation (3.8) is also valid for crystals which, although not possessing a center of symmetry above the Curie point, have at least a symmetry plane perpendicular to the z-direction.

If follows that the state with $P = 0$ is stable for $a>0$ (paraelectric state), while for $a<0$ the stable state must correspond to $P\neq0$ (ferroelectric state).

If one neglects saturation effects, the electric susceptibility above the Curie temperature can be derived by differentiating (3.9) with respect to P and then setting $P = 0$:

$$\chi = P/E = 1/a \ . \tag{3.11}$$

Guided by the experiments, as reflected by Eq. (3.2), it is assumed in Devonshire's theory[5], that around the Curie point the coefficient a can be approximated by a linear function of temperature:

$$a = \alpha(T - T_0) \ . \tag{3.12}$$

We shall see in Sects. 3.4, 5 that moderate variation of a with temperature can be accounted for by anharmonic interactions of thermal vibrations in the lattice. On the other hand, b and c can be considered as temperature independent in a first approximation. From the last two equations one obtains for the inverse susceptibility

$$\chi^{-1} = \alpha(T - T_0) \ . \tag{3.13}$$

Comparison of (3.13) with (3.2) shows that $\alpha = 1/C$, where C is the Curie constant. It is noted that the coefficient a changes sign at the temperature T_0; T_0 is usually near to but not necessarily coincident with the Curie temperature. With (3.12) the following expression is obtained for the free energy:

$$F = F_0 + \tfrac{1}{2}\alpha(T - T_0)P^2 + \tfrac{1}{4}bP^4 + \tfrac{1}{6}cP^6 \ . \tag{3.14}$$

On the other hand, (3.9) can be written in the form

$$E = \alpha(T - T_0)P + bP^3 + cP^5 \ . \tag{3.15}$$

Of the three coefficients appearing in (3.14, 15), α and c are found to be positive in all known ferroelectrics, but b can be either positive or negative. The dependence of the free energy F on temperature and polarization depends very distinctly on the sign of b. It will be shown that if b is negative, (3.14) describes a ferroelectric with a transition of the first order in which the change from the ferroelectric to the paraelectric phase is accompanied by latent heat and a discontinuity in the specific heat. On the other hand, if b is positive, (3.14) describes a material with a transition of the second

---

[5] Devonshire's theory is a special case of Landau's general theory of phase transitions.

order which is accompanid by a peak in the specific heat but not by latent heat. These two cases will be discussed separately.

3.3.2 First-Order Transitions

In (3.14, 15) the parameters α and c are positive and b is assumed to be negative. Setting $E = 0$ in (3.15) and solving this equation for P gives the spontaneous polarization P_s; one obtains either

$$P_s = 0 \tag{3.16a}$$

or[6]

$$P_s = \pm \left(-\frac{b}{2c} \{1 + [1 + 4\alpha c b^{-2}(T_0 - T)]^{1/2}\} \right)^{1/2} . \tag{3.16b}$$

The function $|P_s(T)|$ will be discussed below (Fig. 3.10). At this point we note that (3.16b) also describes the pyroelectric effect, i.e. the change of P_s caused by a change in temperature. Differentiating (3.15) with respect to P gives the inverse susceptibility as a function of the polarization P

$$\chi^{-1} = \alpha(T - T_0) + 3bP^2 + 5cP^4 . \tag{3.17}$$

Above the Curie point and in the absence of large applied fields, the polarization is negligible and (3.17) reduces to (3.13) which represents the Curie-Weiss law. Below the Curie point, however, P is equal to the spontaneous polarization P_s, and the terms in P^2 and P^4 in (3.17) can no longer be neglected. These terms are obtained by substituting for P, in (3.17), the value P_s given by (3.16b):

$$\chi^{-1} = 4\alpha(T_0 - T) + b^2 c^{-1}\{1 + [1 + 4\alpha c b^{-2}(T_0 - T)]^{1/2}\} . \tag{3.18}$$

Equation (3.18) is valid for $T < T_c$, while (3.13) applies for $T > T_c$; the function $\chi^{-1}(T)$ will be discussed below (Fig. 3.11).

The transition from the ferroelectric to the paraelectric phase is illustrated in Fig. 3.9, where $F - F_0$ is shown at different temperatures. Figure 3.9 shows that there are only two minima at $T < T_0$, but at a temperature just above T_0 a third minimum corresponding to $P = 0$ appears. It should be noted that T_0 is the temperature at which the extrapolated χ^{-1} line measured above the Curie point, intersects the temperature axis, see (3.13) and Fig. 3.11. The new central minimum is still less deep than the two outside minima, indicating that the non-polar phase can only be metastable. As the temperature increases, however, the central minimum becomes deeper until at $T = T_c$, it has the same depth as the other two. This temperature

---

[6] Other solutions have been discarded because they correspond to values of P at which the free energy has a maximum rather than a minimum as required for equilibrium.

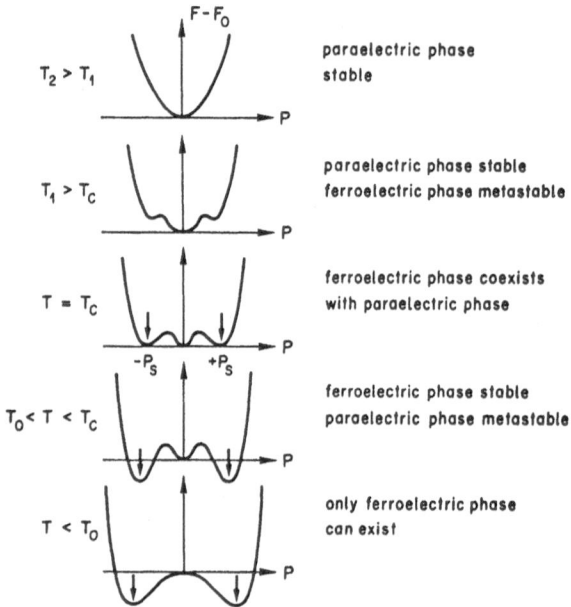

$T_2 > T_1$ — paraelectric phase stable

$T_1 > T_c$ — paraelectric phase stable / ferroelectric phase metastable

$T = T_c$ — ferroelectric phase coexists with paraelectric phase

$T_0 < T < T_c$ — ferroelectric phase stable / paraelectric phase metastable

$T < T_0$ — only ferroelectric phase can exist

Fig. 3.9. Free energy functions, for a ferroelectric with a first-order transition, in different temperature ranges

T_c is called the Curie temperature. At a temperature $T_1 > T_c$ the central maximum is deeper than the outside maxima, indicating that the nonpolar phase is now stable, while the ferroelectric phase is metastable. Finally, at a sufficiently high temperature $T_2 > T_1$ only the paraelectric phase is stable. Note also that below T_c the positions of the absolute minima is at larger values of P_s with decreasing temperature. As T passes through T_c there is a discontinuous change in the position of the absolute minimum.

The Curie temperature T_c and the corresponding value $P_{sc} = P_s(T_c)$ of the spontaneous polarization can be calculated by using the fact that at $T = T_c$ the free energies of the ferroelectric and paraelectric phase are equal, that is

$$F(T_c) = F_0(T_c) \ . \tag{3.19}$$

Using this condition together with (3.14) it follows that

$$\tfrac{1}{2}\alpha(T_c - T_0)P_{sc}^2 + \tfrac{1}{4}bP_{sc}^4 + \tfrac{1}{6}cP_{sc}^6 = 0 \ . \tag{3.20}$$

At the same time the applied field E is set equal to zero in (3.15), which leads to

$$\alpha(T_c - T_0)P_{sc} + bP_{sc}^3 + cP_{sc}^5 = 0 \ . \tag{3.21}$$

These two equations can be solved for P_{sc}^2 and T_c, giving

$$P_{sc}^2 = -\tfrac{3}{4}b/c ,$$ (3.22)

$$T_c = T_0 + \tfrac{3}{16}b^2/\alpha c .$$ (3.23)

It is important to note that, on heating the crystal through the Curie point T_c, the polarization changes discontinuously from $P_{sc} \neq 0$ (for the polar phase) to zero (for the non-polar phase). This behaviour is characteristic of a first-order phase transition ($b<0$). At the Curie point the crystal suddenly loses all the energy associated with the polarization and this causes it to produce a latent heat, typical for a first-order phase transition. Figure 3.10 shows the spontaneous polarization as a function of temperature calculated on the basis of (3.16b), while Fig. 3.11 illustrates the inverse susceptibility $\chi^{-1}(T)$ as well as $\varepsilon(T) = 1 + 4\pi\chi(T)$, based on (3.13 and 18). For the parameters α, b and c, values have been used which are typical for BaTiO$_3$ [Ref. 3.10, pp. 110–111][7]. Figure 3.11 shows that $\chi^{-1}(T)$ and $\varepsilon(T)$ show discontinuities at $T = T_c$. In fact, from (3.13, 23) it follows that

$$\chi_+^{-1}(T_c) = \frac{3}{16}\frac{b^2}{c} ,$$

and from (3.18, 23)

$$\chi_-^{-1}(T_c) = \frac{3}{4}\frac{b^2}{c} .$$

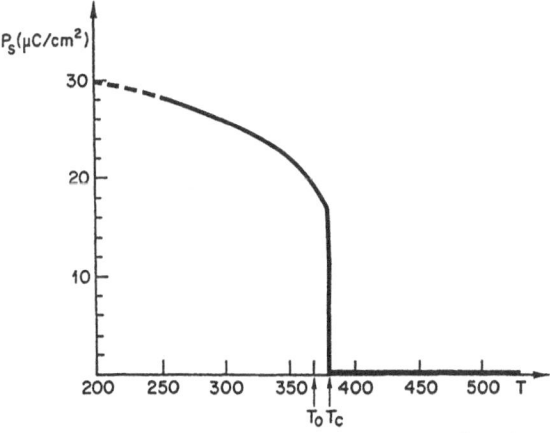

Fig. 3.10. Spontaneous polarization P_s as a function of temperature for a first-order transition, calculated on the basis of Eq. (3.16b) with parameters typical for BaTiO$_3$

[7] The small temperature dependence of b has been neglected [Ref. 3.10, pp. 110–111].

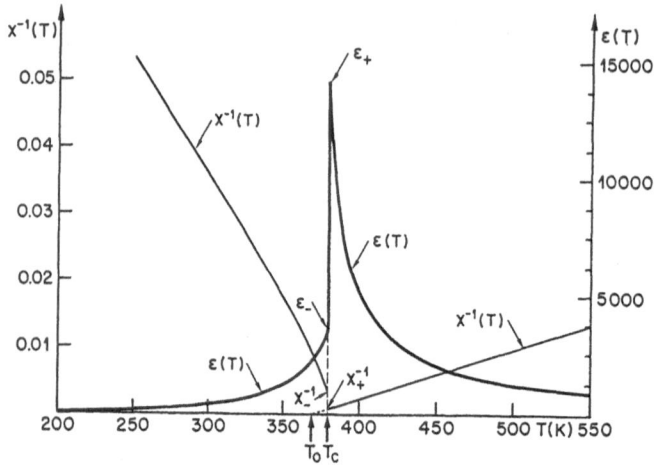

Fig. 3.11. Inverse susceptibility $\chi^{-1}(T)$ and dielectric constant $\varepsilon(T)$ for a first-order transition, calculated on the basis of equations (3.13, 18) with parameters typical for BaTiO₃

From $\varepsilon = 1 + 4\pi\chi$ we obtain correspondingly

$$\varepsilon_+ \cong \frac{64\pi}{3}\frac{c}{b^2} \quad \text{and} \quad \varepsilon_- \cong \frac{16\pi}{3}\frac{c}{b^2} = \frac{1}{4}\varepsilon_+ \ . \tag{3.24}$$

At the transition point T_c the static dielectric constant jumps by a factor of 4.

The discontinuous change of the polarization which occurs at T_c causes a latent heat and a jump in the entropy S, which are typical of a first-order transition. From (3.6) we have

$$S = -\left(\frac{\partial F}{\partial T}\right)_P \ .$$

Using (3.14) and neglecting the small temperature dependence of b and c, for the change of entropy at the transition point T_c we obtain

$$\Delta S = S_0 - S = \tfrac{1}{2}\alpha P_{sc}^2 = -\tfrac{3}{8}\alpha b/c \ , \tag{3.25a}$$

where S_0 is the entropy for zero polarization and where in the latter equation we have used (3.22); remember that $b<0$ and therefore $S_0>S$ as expected. The latent heat at the Curie point is

$$\Delta Q = T_c\Delta S = \frac{1}{2}\alpha T_c P_{sc}^2 = -\frac{3}{8}\frac{\alpha b}{c}T_c \ . \tag{3.25b}$$

ΔS is the increase in entropy associated with the loss of polarization at T_c, and ΔQ is the energy required to destroy the polarization of the crystal.

40

3.3.3 Second-Order Transitions

For a second-order transition all three coefficients α, b and c in (3.14) are positive. With this new assumption, the function $F(P)$, derived form (3.14), has two minima if $T<T_0$ and only one if $T>T_0$, as shown in Fig. 3.12. It is seen from this figure that below T_0 the polarization can have only one of the two values $+P_s$ or $-P_s$, while above T_0 the polarization is zero, from which it follows that $T_0 = T_c$ is the Curie point and according to (3.12)

$$a = \alpha(T - T_c) \tag{3.26}$$

for a second-order transition. The spontaneous polarization is obtained by setting $E = 0$ in (3.15) and accepting only real solutions. For $T \geq T_c$ the only real root of (3.15) is

$$P_s = 0 \ , \tag{3.27a}$$

because α, b and c are positive. For $T<T_c$ there are two real solutions given by

$$P_s = \pm\left(\frac{b}{2c}\{[1 + 4\alpha c b^{-2}(T_c - T)]^{1/2} - 1\}\right)^{1/2} \ , \tag{3.27b}$$

(Fig. 3.13). The phase transition is of second-order because the polarization (3.27b) goes continuously to zero at T_c. The tangent at $T = T_c$ is vertical. Such a behaviour is found in TGS (Fig. 3.3). Before proceeding with the calculation of the dielectric properties, it will be noted that the term in P^5 in (3.15) and the term in P^4 in (3.17) can usually be neglected in the neighbourhood of T_c. This approximation yields simplified expressions for the field E and the inverse susceptibility χ^{-1}, namely

$$E = \alpha(T - T_c)P + bP^3 \ , \tag{3.28}$$

$$\chi^{-1} = \alpha(T - T_c) + 3bP^2 \ . \tag{3.29}$$

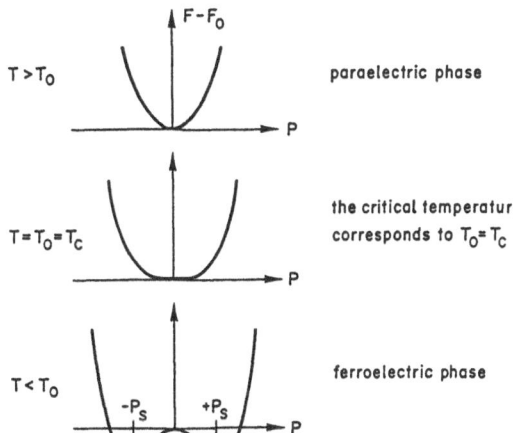

$T>T_0$ paraelectric phase

$T=T_0=T_c$ the critical temperature corresponds to $T_0 = T_c$

$T<T_0$ ferroelectric phase

Fig. 3.12. Free energy functions, for a ferroelectric with a second-order transition, in different temperature ranges

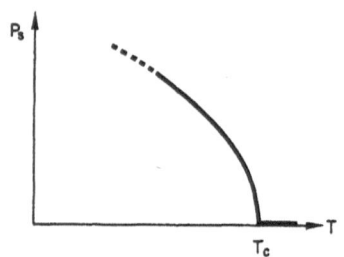

Fig. 3.13. Spontaneous polarization P_s as a function of temperature for a second-order phase transition

Using (3.28) we can derive a simpler form for the spontaneous polarization, namely

$$P_s = \pm[\alpha b^{-1}(T_c - T)]^{1/2} \ . \tag{3.30}$$

Now consider the inverse susceptibility given by (3.29). Above T_c, and in the absence of an applied field, P can be set equal to zero, again yielding (3.13) for $\chi^{-1}(T)$ in the paraelectric state. Below the Curie point, $\chi^{-1}(T)$ is found by substituting (3.30) into (3.29), yielding

$$\chi^{-1}(T) = 2\alpha(T_c - T) \ . \tag{3.31}$$

Thus, below T_c the slope of $\chi^{-1}(T)$ is twice as large and of opposite sign than above T_c. Figure 3.14 shows schematically the behaviour of $\chi^{-1}(T)$ and of $\varepsilon(T) = 1 + 4\pi\chi(T)$. In a second-order phase transition the polarization decreases monotonically with increasing temperature, and reaches zero at $T = T_c$ without discontinuous changes. Consequently, no latent heat can be observed but only a discontinuity in the specific heat. Equation (3.25a), previously derived for a first-order transition, is also valid for a second-order transition. Below the Curie point, P equals the spontaneous polarization P_s, which is given by (3.27b), so that (3.25a) becomes

$$S = S_0 - \tfrac{1}{4}\alpha b c^{-1}\{[1 + 4\alpha c b^{-2}(T_c - T)]^{1/2} - 1\} \ . \tag{3.32}$$

At temperatures below and near the Curie point, where (3.30) is valid, (3.32) reduces to

$$\Delta S = S_0 - S = \tfrac{1}{2}\alpha^2 b^{-1}(T_c - T) \ . \tag{3.33}$$

Since the specific heat is calculated from

$$C = T\left(\frac{\partial S}{\partial T}\right) \ ,$$

one obtains from (3.32)

$$\Delta C = C - C_P = \tfrac{1}{2}\alpha^2 b^{-1}T[1 + 4\alpha c b^{-2}(T_c - T)]^{-1/2} \ , \tag{3.34}$$

where $C_P = T(\partial S_0/\partial T)$ is the specific heat at constant polarization and ΔC is the excess specific heat due to the temperature variation of the po-

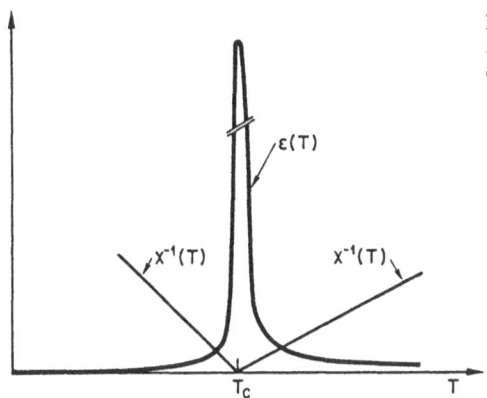

Fig. 3.14. Inverse susceptibility $\chi^{-1}(T)$ and dielectric constant $\varepsilon(T)$ for a second-order transition

larization. It is seen from (3.34) that ΔC is zero at $T = 0$ and increases monotonically with increasing temperature until it reaches the value

$$\Delta C_c = \tfrac{1}{2}a^2 b^{-1} T_c \tag{3.35}$$

at the Curie point T_c. Above the Curie point the polarization is zero, and therefore $\Delta C = 0$.

3.4 Lattice Dynamics of Displacive Ferroelectric Phase Transitions

3.4.1 Ferroelectricity and the LST Relation

A connection between the ferroelectric properties of crystals and lattice dynamics can be seen simply, and perhaps most naturally, through the Lyddane-Sachs-Teller (LST) relation [Ref. 3.1, Sect. 4.3.2]. This connection has already been appreciated by Fröhlich [3.19] in 1949. In its simplest form, in which it is applicable to diatomic ionic crystals of the NaCl or CsCl structures, the LST relation gives the ratio of the static dielectric constant of the crystal to the high frequency dielectric constant in terms of the ratio of the frequencies of the longitudinal optical and transverse optical modes of infinite wavelength $(q = 0)$:

$$\frac{\varepsilon_0}{\varepsilon_\infty} = \frac{\omega_{LO}^2}{\omega_{TO}^2} \ . \tag{3.36}$$

In a ferroelectric crystal, the static dielectric constant follows a Curie-Weiss law temperature dependence above the temperature T_0, (3.1), namely

$$\varepsilon_0 = \varepsilon_\infty + \frac{4\pi C}{T - T_0} \ . \tag{3.37}$$

43

From (3.36) we see that if we exclude the possibility that $\omega_{LO} \to \infty$ as $T \to T_0$, the temperature dependence of ε_0 given by (3.37) implies that ω_{TO} has an anomalous temperature dependence above T_0, of the form

$$\omega_{TO}^2 = \gamma'(T - T_0) \ , \tag{3.38}$$

where γ' is a constant. Inasmuch as negative values of ω_{TO}^2 mean purely imaginary values of the frequency ω_{TO}, and these, in turn, imply normal vibration modes whose amplitudes erupt exponentially with time, the crystal will be unstable below T_0 if it retains the structure it possesses above T_0. In fact, the crystal avoids becoming unstable by transforming to another structure either at or below T_0.

The preceding discussion applies only to diatomic cubic crystals in which each ion has surroundings of at least tetrahedral symmetry. However, as we have seen in [Ref. 3.1, Sect. 4.3.2], it can be extended to crystals possessing more complex structures, which are more interesting for ferroelectric materials. For a diagonally cubic crystal with n atoms in the unit cell (i.e., a crystal in which every atom has surroundings of tetrahedral symmetry, as for example $BaTiO_3$), the generalization of the LST relation is

$$\frac{\varepsilon_0}{\varepsilon_\infty} = \prod_j \frac{\omega_j^2(LO)}{\omega_j^2(TO)} \ . \tag{3.39}$$

In this expression the product extends over the $3n - 3$ optical branches, and $\omega_j(LO)$ and $\omega_j(TO)$ are the infinite wavelength-limiting frequencies of the LO and TO modes, respectively. Again, it is seen that a temperature dependence of ε_0 of the form given by (3.37) implies that one of the TO modes of infinite wavelength in such crystals has an anomalous temperature dependence of the form given by (3.38). Such modes are referred to as *soft modes* in the literature.

We have thus seen that the temperature dependence of ω_{TO}^2 mirrors the temperature dependence of ε_0^{-1}. It should be noted, however, that this qualitative argument is highly oversimplified, since the LST relation is rigorously true only in the harmonic approximation. As we shall see in the following sections, anharmonicity plays a central role in the ferroelectric phase transitions.

3.4.2 Origin of Instability and Polarizability Catastrophe

We have seen that a displacive ferroelectric phase transition is associated with the softening of a TO mode at $q = 0$. In order to obtain more insight into the mechanism of softening, an expression of ω_{TO} in terms of the force constants must be established. For simplicity, we consider a hypothetical diatomic ferroelectric such as an alkali-halide crystal. However, the exercise is not academic since, as discussed at the end of Sect. 3.4, crystals like PbTe and SnTe exhibit a tendency for ferroelectric displacive phase transitions from a rocksalt structure to a rhombohedral structure (incipient ferroelectrics).

Using the shell model discussed in [Ref. 3.1, Sect. 4.3], the frequencies of the TO and LO modes at $q = 0$ are given by

$$\mu\omega_{TO}^2 = S^* - \frac{4\pi e_T^{*2}}{v_a(\varepsilon_\infty + 2)} = S^* - C^* , \tag{3.40a}$$

$$\mu\omega_{LO}^2 = S^* + \frac{8\pi e_T^{*2}}{v_a\varepsilon_\infty(\varepsilon_\infty + 2)} . \tag{3.40b}$$

In these expressions μ is the reduced mass, S^* an effective short range force constant, v_a the volume of the primitive unit cell, ε_∞ the high frequency dielectric constant, and e_T^* is the transverse effective charge defined by

$$e_T^* = \frac{\varepsilon_\infty + 2}{3} z^* e , \tag{3.41}$$

where $z^* e$ is the effective or dynamic charge. The second term, C^*, in (3.40a) represents the force constant from the long-range Coulomb forces. For a real alkali-halide crystal the two force constants S^* and C^* are of the same order of magnitude, but S^* is about twice as great as the other. If they were equal, the crystal would be just unstable for the TO mode at $q = 0$. On the other hand, no cancellation of terms is possible for the LO frequency as may be seen from (3.40b). It is interesting to note that the instability condition for the TO mode may be approached without the crystal becoming unstable for any other mode of vibration [3.20]. From (3.40a) it therefore follows that the instability associated with a ferroelectric phase transition is due to a critical cancellation of short- and long-range forces. Setting $\omega_{TO} = 0$ in (3.40a) and using (3.41), the condition for an instability is given by

$$\frac{4\pi(\varepsilon_\infty + 2)(z^* e)^2}{9v_a S^*} = 1 . \tag{3.42}$$

This condition can be reformulated by introducing the electronic and ionic polarizabilities, α_e and α_i, respectively. The electronic polarizability is given by the Clausius-Mosotti relation [Ref. 3.1, Eq. (4.103)], namely

$$\alpha_e = \frac{3v_a}{4\pi} \frac{\varepsilon_\infty - 1}{\varepsilon_\infty + 2} . \tag{3.43}$$

On the other hand, the dipole moment induced by the relative displacement w of the cation and anion in a TO mode at $q = 0$ is $d_i = z^* ew = \alpha_i E^*$, where α_i is the ionic polarizability and E^* the effective field. In equilibrium the electrostatic force is cancelled by the short-range force: $z^* e E^* = S^* w$, and from these two relations we obtain

$$\alpha_i = \frac{(z^*e)^2}{S^*} . \tag{3.44}$$

Using (3.43, 44) it follows that the instability condition (3.42) can also be written in the form

$$\frac{4\pi}{3v_a}(\alpha_i + \alpha_e) = 1 . \tag{3.45}$$

This is the condition for the *polarizability catastrophe*. The terms "instability" and "polarizability catastrophe" are therefore synonymous.

Until now, our results have been based on the harmonic approximation. It is known however, [Ref. 3.1, Chap. 5] that one effect of the anharmonicity of the lattice vibrations is to make parameters such as S^* linearly temperature dependent. The unit cell volume v_a exhibits a similar dependence (thermal expansion), and on the basis of the shell model, this will also be the case for the dynamic charge z^*e [Ref. 3.1, Eq. (4.84)]. Accordingly we write

$$\frac{\mu\omega_{TO}^2}{S^*} = 1 - \frac{4\pi(\varepsilon_\infty + 2)(z^*e)^2}{9v_a S^*} = \gamma(T - T_0) , \tag{3.46}$$

where γ is a temperature coefficient and T_0 is the temperature at which the crystal will become unstable. For the sake of simplicity, we may think of S^* as being independent of temperature, so that C^* in (3.40a) is responsible for the temperature variation, but this is not a necessary assumption. Equation (3.46) is similar to (3.38) and states that the temperature variation of ω_{TO}^2 is due to anharmonicity. The static dielectric constant is obtained from [Ref. 3.1, Eq. (4.105)]; putting $\omega = 0$ in this equation one obtains

$$\varepsilon_0 = \varepsilon_\infty + \frac{4\pi(\varepsilon_\infty + 2)^2(z^*e)^2}{9v_a \mu\omega_{TO}^2} . \tag{3.47}$$

Using (3.46) we therefore obtain the result

$$\varepsilon_0 = \varepsilon_\infty + \frac{4\pi(\varepsilon_\infty + 2)^2(z^*e)^2}{9v_a S^* \gamma(T - T_0)} . \tag{3.48}$$

Comparing this with the Curie-Weiss law (3.37) we obtain for the Curie constant

$$C = \frac{(\varepsilon_\infty + 2)^2(z^*e)^2}{9v_a S^* \gamma} \simeq \frac{\varepsilon_\infty + 2}{4\pi\gamma} , \tag{3.49}$$

where in the latter equation we have used (3.42).

3.4.3 Soft Modes for Ferroelectric and Antiferroelectric Phase Transitions

It should be noted that the soft mode responsible for a ferroelectric phase transition is a polar mode, that is, it induces a non-vanishing polarization P due to the relative displacements of cations and anions as illustrated schematically in Fig. 3.15a. At the transition point the displacement becomes static which produces a permanent polarization P. The static displacement is limited by anharmonic forces and determines the new structure in the ferroelectric phase. The ions then vibrate around the new equilibrium positions.

An instability may arise for a value of q other than zero. Of course, such a mode is not a polar mode and will therefore not produce a permanent polarization. For example, it can be shown that in certain circumstances a crystal of the caesium chloride type will become unstable against a transverse mode for which q is at the point $(\pi/a)(1,0,0)$ in reciprocal space. In this mode the atoms move as shown in Fig. 3.15b, with one Bravais lattice remaining undistorted. This type of instability results in a transition to an *antiferroelectric* phase [3.21–23]. These deformations, even if they do not give a spontaneous polarization, may be accompanied by changes in the dielectric constant. An example of an antiferrodistortive transition of $SrTiO_3$ is discussed in Sect. 3.6 (Fig. 3.22).

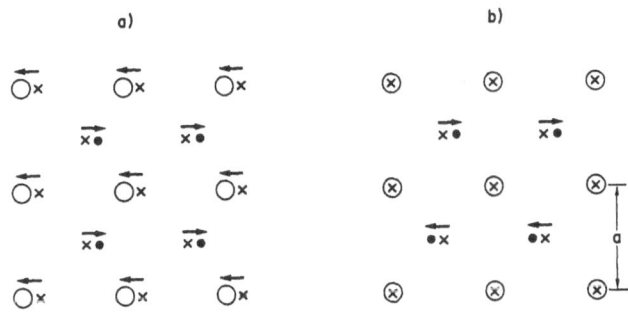

× equilibrium position

Fig. 3.15a,b. Displacive phase transition in two dimensions. (a) Ferroelectric phase: condensation of the TO-mode at $q = 0$, leading to a non-vanishing polarization P. (b) Antiferroelectric phase: condensation of a transverse mode at $q = (\pi/a)(0, 1)$ with $P = 0$. The crosses indicate the equilibrium positions above T_c. *Large open circles*: anions, *small closed circles*: cations. The arrows indicate the displacement of the ions

3.4.4 Lattice-Dynamical Basis of Devonshire's Theory

In the harmonic approximation the equation of motion for the TO mode at $q = 0$ of diatomic cubic crystals is given by

$$\mu\ddot{w} + (S^* - C^*)w = 0 \; , \tag{3.50}$$

where S^* and C^* are defined by (3.40). At the transition temperature $S^* = C^*$ and the equation of motion is now $\ddot{w} = 0$ with the solution $w = at + b$, which implies that the separation between the ions in the primitive unit cell increases indefinitely with time. However, at this point we must modify the analysis to include anharmonic effects, which become important for large w. The simplest way to do this is to rewrite (3.50) in the form

$$\mu\ddot{w} + R^*w + Bw^3 + B'w^5 = 0 \; , \tag{3.51}$$

where $R^* = S^* - C^*$ and B, B' are anharmonic short range force constants. In writing (3.51) we have singled out the TO mode at $q = 0$, and ignored all other modes. However, by introducing anharmonic terms in the restoring force, we have destroyed the normality of the modes which alone makes this procedure admissible. Nevertheless, we use this approximation and the special form of the equation of motion (3.51) to establish the connection with Devonshire's macroscopic theory. The new restoring force $(R^*w + Bw^3 + B'w^5)$ may be integrated with respect to w to give the anharmonic potential Φ governing the dynamics; one obtains

$$\Phi = \tfrac{1}{2}R^*w^2 + \tfrac{1}{4}Bw^4 + \tfrac{1}{6}B'w^6 \; . \tag{3.52}$$

Note that for the sake of simplicity and in order to make direct contact with Devonshire's theory, only even powers of w are included in Φ. The model then consists of a single particle with mass μ and charge z^*e moving in the anharmonic potential Φ. In the following we follow *Cochran* [3.20] but we simplify the analysis by considering only the one-dimensional problem (3.52) instead of the three-dimensional one.

We now consider the behaviour of the crystal in a *static* applied field E. Under the influence of the applied field E the cations and anions will be displaced in opposite directions until a relative static equilibrium displacement w_0 is reached. Since the pattern of ionic displacements induced by the static applied field is the same as in a TO mode for $q = 0$, the effective field acting on the ions is given by [Ref. 3.1, Eq. (4.95)]

$$E^* = E + \frac{4\pi}{3}P \; , \tag{3.53}$$

where P is the static polarization associated with the relative displacement of the two Bravais lattices. From [Ref. 3.1, Eq. (4.11)] we have

$$P = \frac{(\varepsilon_\infty + 2)z^*e}{3v_a}w_0 + \frac{\varepsilon_\infty - 1}{4\pi}E \; . \tag{3.54}$$

The equilibrium displacement of the two Bravais lattices is then determined by the balance between short-range and electrostatic forces, expressed by

48

$$z^* e E^* = S^* w_0 + B w_0^3 + B' w_0^5 \; . \tag{3.55}$$

Using (3.40, 41, 53) Eq. (3.55) can be rewritten in the form

$$[z^* e + \tfrac{1}{3}(\varepsilon_\infty - 1)] E = R^* w_0 + B w_0^3 + B' w_0^5 \; , \tag{3.56}$$

where $R^* = S^* - C^*$, and in the harmonic approximation ($B = B' = 0$) the static displacement increases indefinitely if R^* approaches zero. The lattice is then stabilized only by the anharmonic forces which will limit the relative displacement to a finite value. Substituting (3.53, 54) into (3.55) and using (3.46) gives

$$\frac{\varepsilon_\infty + 2}{3} z^* e E = S^* \gamma (T - T_0) w_0 + B w_0^3 + B' w_0^5 \; .$$

Using (3.54) the latter equation becomes

$$\left[1 + \frac{9 v_a}{4\pi} \frac{(\varepsilon_\infty - 1) S^*}{(\varepsilon_\infty + 2)^2 (z^* e)^2} \gamma (T - T_0) \right] E$$

$$= \frac{9 v_a S^*}{(\varepsilon_\infty + 2)^2 (z^* e)^2} \gamma (T - T_0) P + \frac{3}{(\varepsilon_\infty + 2) z^* e} (B w_0^3 + B' w_0^5) \; .$$

Introducing the Curie constant C from (3.49) we obtain

$$\left[1 + \frac{(\varepsilon_\infty - 1)(T - T_0)}{4\pi C} \right] E = \frac{T - T_0}{C} P + \frac{3}{(\varepsilon_\infty + 2) z^* e} (B w_0^3 + B' w_0^5) \; . \tag{3.57}$$

The term $(\varepsilon_\infty - 1)(T - T_0)/4\pi C$ is less than $(T - T_0)/C$ and also less than $\gamma (T - T_0)$ and from (3.46, 49) it therefore follows that this term can be neglected compared to unity. To about the same accuracy we can now express w_0 in terms of P in (3.57) by using (3.54) thereby neglecting the second term in (3.54). We than obtain the approximate result

$$E = \frac{T - T_0}{C} P + v_a^3 g^4 B P^3 + v_a^5 g^6 B' P^5 \; , \tag{3.58}$$

where we have introduced the abbreviation

$$g = \frac{3}{(\varepsilon_\infty + 2) z^* e} \; . \tag{3.59}$$

Using (3.6 and 58) we obtain for the free energy per unit volume at constant temperature

$$F = F_0 + \frac{T - T_0}{2C} P^2 + \frac{1}{4} v_a^3 g^4 B P^4 + \frac{1}{6} v_a^5 g^6 B' P^6 \; . \tag{3.60}$$

49

Comparing (3.60) with (3.8) and (3.58) with (3.15) one obtains the following expressions of the phenomenological constants a, α, b and c of Devonshire's model in terms of the microscopic parameters

$$a = \frac{T - T_0}{C} \quad , \quad \alpha = 1/C \quad , \tag{3.61a}$$

$$b = v_a^3 g^4 B \quad , \quad c = v_a^5 g^5 B' \quad . \tag{3.61b}$$

We are now able to discuss the transitions in terms of atomic rather than macroscopic parameters. We confine ourselves to a discussion of the results (Problem 3.9.1). For a first-order transition with $B<0$ it may be shown that if T_c is approached from above, the frequency at T_c is given by

$$\mu \omega_{TO}^2(T_c+) = \frac{3B^2}{16B'} \quad , \tag{3.62}$$

and that the relative "jump" of the ions at T_c, in going to the ferroelectric phase, is

$$w_0(T_c) = w_c = \left(\frac{3|B|}{4B'}\right)^{1/2} \quad . \tag{3.63}$$

The spontaneous polarization at T_c follows from (3.22,61b) and is given by

$$P_{sc} = \frac{w_c}{g v_a} \quad . \tag{3.64}$$

The Curie constant C is given by (3.49), while from (3.23,61) we obtain

$$T_c - T_0 = \frac{3B^2 C v_a g^2}{16B'} \quad . \tag{3.65}$$

The maximum value of the static dielectric constant is approximately

$$\varepsilon_+ = \frac{64\pi B'}{3g^2 v_a |B|} \tag{3.66}$$

just before the transition, while $\varepsilon_- = \varepsilon_+/4$ is the dielectric constant just after the transition, and $\varepsilon_0(T_0) = 3\varepsilon_+/32$.

In the ferroelectric phase the ions vibrate about the displaced positions with frequencies determined by the anharmonic force constants for T close to T_c:

$$\mu \omega_{TO}^2(T_c-) = \frac{3B^2}{4B'} \tag{3.67a}$$

just after the transition, and

$$\mu \omega_{TO}^2(T_0) = 2\frac{B^2}{B'} \quad . \tag{3.67b}$$

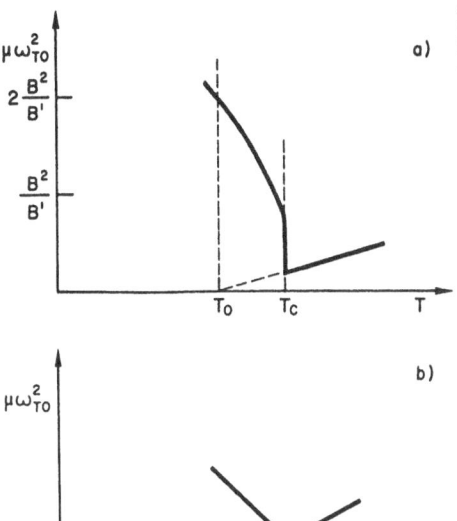

Fig. 3.16a,b. $\mu\omega_{TO}^2$ as a function of temperature for (a) a first-order transition, (b) a second-order transition

Figure 3.16a shows the temperature dependence of $\mu\omega_{TO}^2$. Note the jump at T_c and the different slopes above and below T_c. If the temperature dependence of ω_{TO}^2, as shown in Fig. 3.16a, is used in the LST relation (3.36) we obtain the behaviour of $\varepsilon_0(T)$ as illustrated in Fig. 3.11.

From (3.25,61) one finds for the entropy change ΔS_c and the latent heat ΔQ_c at T_c:

$$\Delta S_c = S_0 - S = -\frac{3}{8}\frac{S^*}{v_a}\gamma\frac{B}{B'} > 0 \;, \quad \text{and} \tag{3.68a}$$

$$\Delta Q_c = -\frac{3}{8}\frac{S^*}{v_a}\gamma\frac{B}{B'}T_c \;. \tag{3.68b}$$

For a second-order transition with $B>0$ one finds (Problem 3.9.1)

$$\mu\omega_{TO}^2(T \geq T_c) \cong S^*\gamma(T - T_c) \tag{3.69a}$$

$$\mu\omega_{TO}^2(T \leq T_c) \cong 2S^*\gamma(T_c - T) \;. \tag{3.69b}$$

Figure 3.16b shows $\mu\omega_{TO}^2$ as a function of temperature. Together with the LST relation (3.36) the function $\omega_{TO}^2(T)$ shown in Fig. 3.16b mirrors the temperature dependence of ε_0 as illustrated in Fig. 3.14. In principle, the minimum value of ω_{TO} is now zero at T_c, just as the maximum value of ε_0 is infinite. Near T_c the linear term in the short-range force will almost be cancelled by the electrostatic force and consequently the vibration will be very anharmonic.

51

Finally, the excess specific heat due to the temperature variation of the polarization is obtained from (3.35, 61) and is given by

$$\Delta C_c = \frac{1}{2} \frac{(S^* \gamma)^2}{B v_a} T_c$$

(3.70)

at the Curie temperature.

3.5 Lattice Dynamical Models for Ferroelectric Phase Transitions

In Sect. 3.4.4 we have accounted for anharmonicity only insofar as we have considered the motion of the ions in the TO mode at $q = 0$ to be that of a single particle in an anharmonic potential. In doing so we have completely neglected the interaction between different phonons. In Sect. 3.5.1 we discuss the application of the anharmonic lattice model to ferroelectric phase transitions; this approach is more basic because it accounts explicitly for anharmonic interactions. Some of the results previously discussed will be recovered, and a brief discussion will also be given to the *response function* and the *central mode* of ferroelectric crystals. In Sect. 3.5.2 we discuss the electronic theory of ferroelectricity which forms the basis of a microscopic description of the softening mechanism, and in Sect. 3.5.3 we discuss some results obtained from the polarizability model which has recently been developed and applied with success to some ferroelectrics. The discussion will be qualitative; we shall only quote some results and discuss their physical significance.

3.5.1 The Anharmonic Lattice Model

In Sect. 3.4.2 we have seen that a crystal can become unstable due to a cancellation or overcancellation between short-range and long-range forces. If $\omega_j(q = 0) = \omega_j(0)$ is the harmonic frequency of the soft mode, the instability implies that $\omega_j^2(0) \leq 0$ at all temperatures. In [Ref. 3.1, Sects. 5.5.2,3] we saw, however, that anharmonic interactions make the effective frequency of any mode a function of temperature and of the frequency with which the crystal is probed. As an approximation we can neglect this latter complication and introduce the pseudoharmonic frequency $\tilde{\omega}_j(0)$, which depends only on temperature and which is the frequency determined experimentally by spectroscopic techniques. According to [Ref. 3.1, Eq. (5.144)], $\tilde{\omega}_j(0)$ is defined by

$$\tilde{\omega}_j^2(0) = \omega_j^2(0) + 2\omega_j(0)\Delta_j(0, \tilde{\omega}_j(0)) \ ,$$

(3.71)

where $\Delta_j(0, \tilde{\omega}_j(0))$ is the real part of the temperature dependent self-energy defined by [Ref. 3.1, Eq. (5.155)]. It is then possible to have $\omega_j^2(0)<0$, but $\tilde{\omega}_j^2(0)>0$. As we have seen in [Ref. 3.1, Sect. 5.5.3, Eqs. (5.155–158)] the temperature dependence of the shift in frequency is very complicated, but at relatively high temperatures the difference between $\tilde{\omega}_j^2(0)$ and $\omega_j^2(0)$ is predicted to be proportional to temperature. If we write the temperature dependent part [the second term in (3.71)] as $\gamma'T$, where γ' is a positive constant which is defined by this identification, and if we put $\omega_j^2(0) = -\gamma'T_0$, which is equivalent to determining $\omega_j^2(0)$ in terms of the known values of T_0 and γ', we can rewrite (3.71) in the form

$$\tilde{\omega}_j^2(0) = \gamma'(T - T_0) \; , \tag{3.72}$$

which is identical with (3.38) obtained from the LST relation and the Curie-Weiss law. It should be noted that according to [Ref. 3.1, Eqs. (5.155–158)] the self-energy and hence γ' depend not only on the quartic but also on the cubic anharmonic coupling coefficients and contain the effect of thermal expansion [3.24].

In [Ref. 3.1, Sect. 5.5.2] we have also discussed the response function, for which we have found the following approximate expression [Ref. 3.1, Eq. (5.145)]

$$R_j(\boldsymbol{q}, \omega) = [\tilde{\omega}_j^2(\boldsymbol{q}) - \omega^2 - 2i\tilde{\Gamma}_j(\boldsymbol{q})\omega]^{-1} \; , \tag{3.73}$$

where $\tilde{\Gamma}_j(\boldsymbol{q})$ is the damping constant. If we consider the response of the crystal to light in the infrared or far-infrared region, the complex dielectric constant is approximately given by [Ref. 3.1, Eq. (5.136)]

$$\varepsilon(\omega) = \varepsilon_1(\omega) + i\varepsilon_2(\omega) \cong \varepsilon_\infty + \sum_j S_j R_j(0, \omega) \; ,$$

where S_j is the oscillator strength of the mode j. The contribution of the soft mode $(0j)$ to $\varepsilon_2(\omega)$ is therefore given by

$$\varepsilon_{2j}(\omega) = \frac{2S_j\tilde{\Gamma}_j(0)\omega}{[\tilde{\omega}_j^2(0) - \omega^2]^2 + 4\omega^2\tilde{\Gamma}_j^2(0)} \; . \tag{3.74}$$

For $\tilde{\Gamma}_j^2(0) \ll \tilde{\omega}_j^2(0)$, $\varepsilon_{2j}(\omega)$ describes a resonant behavior with a peak at $\tilde{\omega}_j(0)$ and line width $2\tilde{\Gamma}_j(0)$. Such a behavior is found in SrTiO$_3$ for temperatures well above the extrapolated transition temperature $T_c \cong 32\,\mathrm{K}$, as illustrated in Fig. 3.17 for the lowest frequency $q = O$ TO mode [3.25]. If, however, $2\tilde{\Gamma}_j^2(0) \geq \tilde{\omega}_j^2(0)$, Eq. (3.74) shows that the spectrum has no maximum at or near $\tilde{\omega}_j(0)$, but only at $\omega = 0$. The mode is then overdamped and is called a *central mode*. This situation seems to apply for certain modes in the lowest TO branch of BaTiO$_3$. The most probable origin of the large damping is the onset of a partial disorder near T_c. In simple terms this means

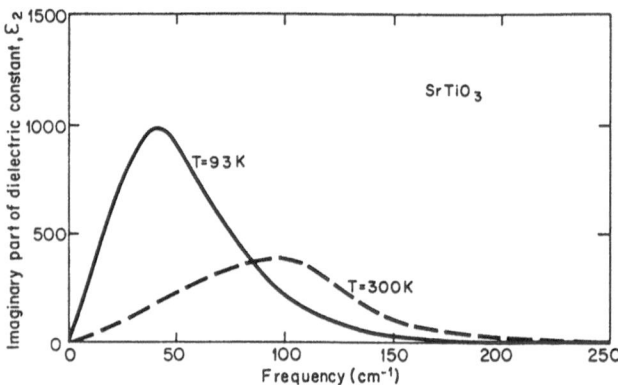

Fig. 3.17. The imaginary part of the dielectric constant of $SrTiO_3$ as a function of frequency in the far infrared region, for two different temperatures [3.25]

that the particles involved in the soft mode move near T_c in a double (or multi)-minimum potential with a potential barrier between minima of the order of $k_B T_c$, the distance between the minima being of the order of 0.1 Å. The transition is then not purely displacive but also contains a component of the order-disorder type (relaxational mode). The associated soft-mode spectrum consists of at least two peaks, the higher frequency corresponding to the small-amplitude oscillations in a single minimum (classical resonant soft-mode), and the lower frequency corresponding to the large-amplitude hopping among different minima (relaxational motion), which is strongly anharmonic and gives rise to the central peak [3.26].

Other anharmonic lattice models for ferroelectric phase transitions have been studied by *Gillis* [3.27], *Thomas* [3.28], *Silverman* [3.29], and by *Pytte* [3.30]. A good review about the spectroscopy of soft modes has been given by *Scott* [3.31].

3.5.2 Electronic Theory of Soft-Mode Instability

Given the fact that in the harmonic approximation $\omega_j^2(0) < 0$ for the soft mode, in the preceding section we have shown how the anharmonic terms in the crystal potential energy stabilize this mode by renormalizing its frequency to real values. This treatment was however formal and did not address the question as to the microscopic origin of the cancellation between the Coulomb and short-range contributions. Why does such a cancellation or near-cancellation exist in crystals such as $BaTiO_3$, $SrTiO_3$ or PbTe and SnTe, but not for instance in crystals such as the alkali halides or alkaline-earth fluorides? This question can only be answered from first-principle microscopic theories, in which the force constants are derived from the electronic structure of the crystal.

Within the harmonic approximation several first-principle [3.32, 33] and parameter-free approaches [3.34] have been recently developed which may be used to investigate ferroelectricity from a microscopic point of view. In the following we give a brief discussion of the results of the pseudopotential method used to develop a simple model for zone-center phonon modes and ferroelectricity [3.35]. This model has been applied specifically to the IV–VI compounds which we have discussed at the end of Sect. 3.2. Using this model the harmonic part of the zone center TO phonon mode ω_{TO} for cubic, binary crystals has been evaluated and the following result obtained

$$\omega_{TO}^2 = \frac{e^2}{\mu v_a}(I - E) \ . \tag{3.75}$$

Here μ and v_a are the reduced mass and the volume of the primitive unit cell, respectively. The term I in (3.75) originates from the short-range $q = 0$ part of the ion-ion interaction. It simply describes the plasma vibration of the positive ion cores embedded in a homogeneous negative background. On the other hand, the electronic part E depends on the structure of the valence band and tends to reduce or "screen" the plasma vibration frequency of the ion cores. Typically, one finds $I = 70$ and $E = 30$ in covalent semiconductors such as Si or GaAs, while $(I - E)/I < 10^{-3}$ in IV–VI compounds [3.36].

From a detailed analysis it follows that $I - E$ is small and hence the crystal can be expected to show a soft mode, if

i)   its structure is ionic (NaCl or CsCl structure), and
ii)   the weak electronegativity of its constituent ions favours already covalent bonding.

If these two conditions are satisfied E will be large. In PbTe, for example, which has the NaCl structure, the covalency is comparable to the III–V compounds. The III–VII thallous halides represent another class of materials which fulfill both conditions (i) and (ii) and are indeed known to have quasi-soft, i.e. strongly temperature dependent TO phonon frequencies [3.37] [Ref. 3.1, Fig. 5.14]. In these structures, therefore, ionic structure together with covalency tend to produce the ferroelectric instability. It should be mentioned, however, that in many of the "classical" ferroelectrics, such as the perovskites, the strong anisotropy of the local field is also known to be essential [3.38]. Further investigations are needed to clarify the quantitative role of covalency in these materials; doubtless it is a supporting element for the ferroelectric instability. Ferroelectric semiconductors have been discussed in the book by *Fridkin* [3.39].

3.5.3 The Polarizability Model

Experience has shown that more than 98% of all crystals which exhibit a soft-mode behavior, contain either oxygen or other chalcogenide ions. This fact suggests a central role of chalcogenide ions in ferroelectric phase tran-

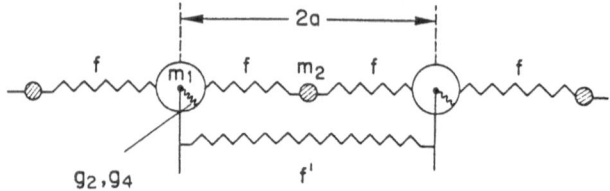

Fig. 3.18. Diatomic linear chain model for a ferroelectric [3.43]

sitions. For perovskites [3.40], SbSI [3.41] and K_2SeO_4 [3.42] it has been shown that the quartic oxygen ion polarizability governs their soft-mode behavior. The "pathological" behavior of O^{2-} and its homologues has been taken into account in a linear diatomic shell model by *Bilz* et al. [3.43]. While the cation is assumed to be rigid, the anion core-shell coupling consists of a strongly attractive harmonic electron-ion coupling constant g_2 and a non-linear quartic repulsive term g_4 (Fig. 3.18).

The linear diatomic shell model mirrors some essential features of three-dimensional incipient ferroelectrics such as PbTe and SnTe. These compounds show a tendency for ferroelectric displacive phase transitions from a high-temperature rocksalt structure to a rhombohedral phase at low temperatures, the static displacements of the ions being in the (111) direction. Since planes perpendicular to the (111) direction of the NaCl-structure contain only one type of ion, each ion in Fig. 3.18 actually represents a whole layer of ions and each displacement dipole represents a whole layer. Since dipole-dipole interactions between layers of dipoles are known to be short-range [3.44], it is sufficient to introduce short-range constants only, as shown in Fig. 3.18. The instability of the ferroelectric soft mode is attributed to the attractive Coulomb forces acting between the anion and its shell which is modeled by the force constant $g_2 < 0$. The stabilization in the paraelectric and ferroelectric phases is achieved via the on-site fourth-order core-shell coupling constant $g_4 > 0$. The relative core-shell motion of each anion l is therefore that of a particle moving in a double minimum potential of the form $V = g_2 w^2(l) + g_4 w^4(l)$, where $w(l)$ is the relative displacement between an anion core and its shell; accordingly the order parameter of this model is the thermal average $\langle w(l) \rangle_T$, rather than the relative cation-anion displacement.

If $u_\kappa(l)$ and $v(l)$ are the displacements of cores and anion shell in unit cell l, respectively, then the equations of motion are written in the form

$$m_1 \ddot{u}_1(l) = g_2 w(l) + g_4 w^3(l) + f'[u_1(l+1) + u_1(l-1) - 2u_1(l)]$$
$$m_2 \ddot{u}_2(l) = f[v(l) + v(l+1) - 2u_2(l)]$$
$$0 = -g_2 w(l) - g_4 w^3(l) + f[u_2(l) + u_2(l-1) - 2v(l)] , \qquad (3.76)$$

where $w(l) = v(l) - u_1(l)$.

For the temperature dependence of the soft mode it is necessary to study the approximate solutions in the self-consistent phonon approximation (SPA). The SPA corresponds to a linearization of the cubic terms which enter via g_4 in the equations of motion (3.76):

$$g_4 w^3(l) \cong 3g_4 \langle w^2(l) \rangle_T w(l) , \quad \text{where} \tag{3.77a}$$

$$\langle w^2(l) \rangle_T = \frac{\hbar}{2m_1 N} \sum_{qj} \frac{1}{\omega_j(q)} coth \frac{\hbar\omega_j(q)}{2k_B T} . \tag{3.77b}$$

In (3.77) the quantity $\langle w^2(l) \rangle_T$ is the self-consistent thermal average of $w^2(l)$ at temperature T. For an evaluation of $\langle w^2(l) \rangle_T$ on the basis of a three-dimensional isotropic Debye model the reader is referred to the literature [3.45]. Here we simply quote the result obtained for the frequency of the $q = 0$ ferroelectric soft mode [3.45]. If (3.77) is substituted in (3.76) the shell coordinate $v(l)$ can be liminated and for the frequency of the soft mode one obtains

$$\omega_f^2 = \omega_0^2 \frac{g(T)}{2f + g(T)} , \quad \text{where} \tag{3.78}$$

$$g(T) = g_2 + 3g_4 \langle w^2(l) \rangle_T \tag{3.79}$$

and $\omega_0^2 = 2f/\mu$ is the rigid-ion limit of the ferroelectric soft mode for $g(T) \rightarrow \infty$; μ is the reduced mass. In the paraelectric phase $(T > T_c)$, $g(T)$ is positive, while in the ferroelectric phase $g(T)$ is replaced by $-2g(T)$, that means it behaves like the Landau parameter $\chi^{-1}(T)$ defined by (3.29, 31). At the phase transition $g(T_c) = 0$ and hence $\omega_f^2 = 0$. It is also found that for $k_B T \gg \hbar\omega_f$, $\hbar\omega_D$, and for $\omega_f^2 < \omega_0^2$, $\omega_f^2 \sim (T - T_c)$ as in (3.69a). It should be noted that since $g(T)$ depends on ω_f^2 and T via (3.77b), the substitution of (3.79) into (3.78) yields a single self-consistent equation which gives the implicit relation between ω_f^2 and T. The results of this model have been applied to PbS, PbSe, PbTe, SnTe [3.45] as well as to $KTaO_3$ [3.43] and $SrTiO_3$ [3.46]. In the latter case the TiO_6 octahedron is treated as a polarizable anion and the parameter g_2 describes not only the polarizability of the oxygen ion, but in addition the attractive (negative) Ti-O Coulomb interaction which, in fact, is responsible for the instability of the ferroelectric soft mode in the harmonic approximation.

3.5.4 Range of Validity of the Landau-Devonshire Theory (By U.T. Höchli)

The lattice-dynamical basis for the Landau-Devonshire theory involves softening of TO modes with $q = 0$. In such a mode, all displacements are in phase, and accordingly, the polarization carried by this mode is homogeneous. Anharmonicity couples modes with different q and as a consequence the macroscopic polarization fluctuates in space. The lowest-order term in the Landau free energy to account for fluctuation is

$$\Delta F_{\mathrm{Fluct}} = \tilde{c}(\mathrm{grad}\ P)^2 \ . \tag{3.80}$$

A statistical-mechanics treatment [Ref. 3.2a, p. 371] of the local polarization $P(r)$ leads to a correlated polarization of the form

$$\langle P(r) \cdot P(r') \rangle = \exp\left(-|r - r'|/\xi\right) \frac{kT}{|r - r'|8\pi\tilde{c}} \ , \tag{3.81a}$$

where

$$\xi = [2\tilde{c}/\alpha(T - T_c)]^{1/2} \ , \tag{3.81b}$$

$\langle A \rangle$ denotes the statistical average of the quantity A and α is defined by (3.14). From (3.81b) it follows that the correlation length itself is a critical function of temperature; it diverges at T_c. Ginzburg argued that the Landau-Devonshire approximation (L–D) is valid at most if the polarization fluctuations averaged within the correlation volume ξ^3 remained small compared to $\langle P^2 \rangle$. He showed that this criterion led to an exclusion of (L–D) out of a temperature range given by

$$\varepsilon \equiv \frac{|T_c - T|}{T_c} = f(k_B b / \tilde{c}\alpha)\lambda^2 \ , \tag{3.82}$$

where λ is the correlation length at zero temperature, f a numerical factor of the order of 100 and the constants α and b are defined by (3.12–14). This correlation length is an indication for the range of forces between lattice constituents. The range of exclusion ε for L–D thus goes as λ^2 (not as λ^{-6} as generally stated: The λ^{-6} dependence of ε is an artifact stemming from the introduction of the λ-dependent specific heat into these considerations). The fact that the Curie-Weiss law is so universally observed is thus attributed quite generally to the long range of electric-type interactions and the concomitant smallness of the critical range ε. However, it has been pointed out that dipolar interaction is not truly long-ranged and, in particular, that the critical susceptibility for a 3d dipolar system was given by [3.47]

$$\varepsilon \sim (T - T_c)\gamma_{\mathrm{dip}} \ , \tag{3.83}$$

where $\gamma_{\mathrm{dip}} = 1.033$. This value is hard to distinguish from the L–D value $\gamma = 1$. The failure to observe critical behaviour in a ferroelectric thus could well be due to the small distinction between the L–D and the critical parameters and not to the vanishing L–D region.

3.6 Soft-Mode Spectroscopy of Selected Ferroelectrics

In the following we discuss the temperature dependence of the soft-mode frequency of some ferroelectric and antiferroelectric crystals, as observed by means of various experimental techniques.

Figure 3.19 shows a schematic plot of the temperature dependence of the TO soft mode of $BaTiO_3$. The full lines are data measured by *Luspin* et al. [3.48] in the cubic phase by using infrared spectroscopy, and in the tetragonal phase by *Burns* [3.49] and *Scalabrin* et al. [3.50]. The broken lines represent the mode frequencies in the two other phases (Fig. 3.6), estimated from Raman scattering experiments [3.51]. We first note that the qualitative behaviour of $\omega_{TO}(T)$ is similar to that shown in Fig. 3.16a for a first-order transition; note, however, that in the latter figure we have plotted $\omega_{TO}^2(T)$, rather than $\omega_{TO}(T)$. In the cubic paraelectric phase the soft mode is a tryply degenerate mode of species F_{1u} which in the tetragonal phase splits into a doubly degenerate E mode and a non-degenerate A_1 mode. Irrespective of the first-order character of the phase transition, the E-type soft mode component continues to soften on cooling and appears to be driven by a lower-temperature phase transition. It is only the mode of species A_1 which shows an abrupt discontinuity at the cubic-tetragonal phase transition, and it is the B_1 mode which changes abruptly at the tetragonal-orthorhombic phase transition, while the B_2 mode continues to soften in the orthorhombic phase. The softening of the E-mode in the tetragonal phase explains the

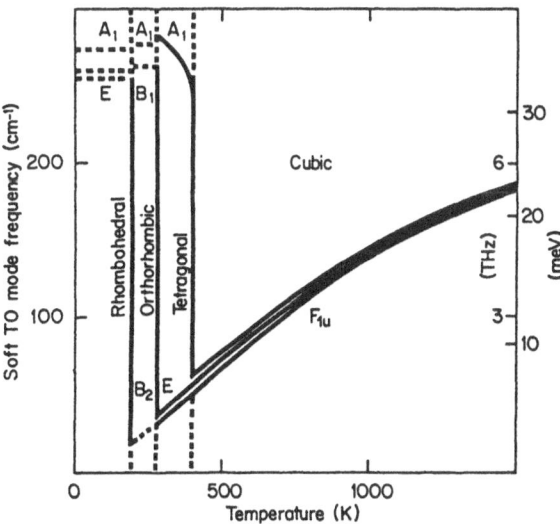

Fig. 3.19. A schematic plot of the observed temperature dependence of the three components of the TO soft mode for $BaTiO_3$ [3.48–51]

departure from the observed Curie law in the static dielectric constant at the approach of T_c in the cubic phase, owing to the weight of this E-component in the F_{1u}-type mode. This discrepancy might be due to the onset of a dynamical disorder which leads to a central mode in addition to the soft mode as discussed in Sect. 3.5.1.

The phonon dispersion curves of SrTiO$_3$ have been determined by neutron scattering [3.52] and are shown in Fig. 3.20 for q parallel to (100) and to (111) directions. There are five atoms per unit cell (Fig. 3.5), so that for a general value of q there are 15 branches, i.e., values of j. However, for q in the directions just mentioned, pairs of transverse branches have the same frequency giving in effect five transverse and five longitudinal branches, not all of which are shown in Fig. 3.20. It will be noticed that for values of q close to the origin there is a dip in the frequency of the lowest TO branch at room temperature, and this dip becomes more pronounced at 90 K. Figure 3.21 shows the temperature dependence of the soft mode [3.53,54]. This mode is infrared active and produces a resonance in $\varepsilon_2(\omega)$ as shown in Fig. 3.17.

From Fig. 3.21a it is seen that ω_{TO}^2 at $q = 0$ is quite accurately linear in T over the range shown. The same diagram shows $\varepsilon_0^{-1}(T)$, which is also found to vary as $T - T_0$ in this range, with the same value $T_0 \cong 30$ K as fits (3.72). However, the crystal does not make a transition to a ferroelectric phase; below about 60 K the dielectric constant increases less rapidly than predicted by (3.1) and tends to a constant value as $T = 0$ K is approached. Evidently the mode we are considering remains stable and this has been confirmed by measurements of its frequency at low temperatures shown in

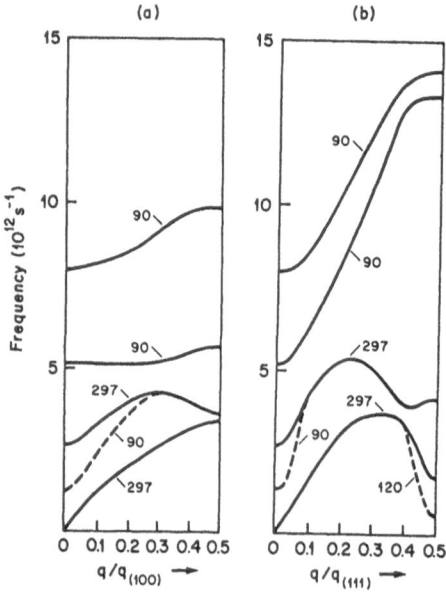

Fig. 3.20a,b. Experimentally measured frequencies $\tilde{\omega}_j(q)/2\pi$ for SrTiO$_3$. In (a) q is parallel to (100), in (b) to (111). Only transverse modes are shown. A number such as 90 indicates the measurements were made at 90 K; where frequencies vary considerably with temperature a branch at lower temperature is shown by (- - -) [3.52]

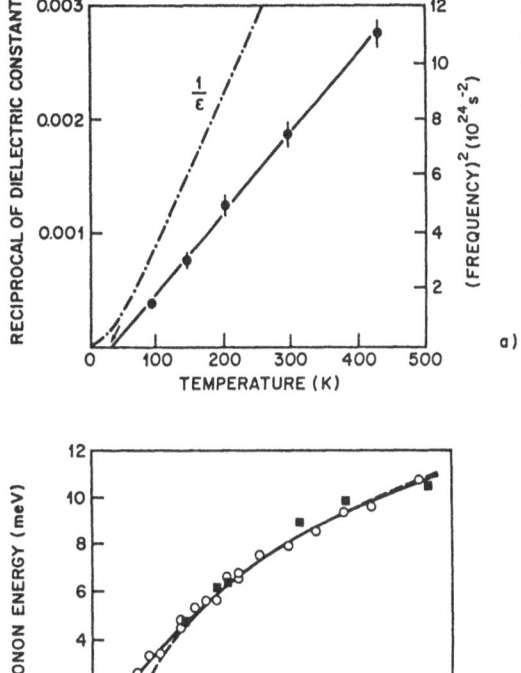

Fig. 3.21. (a) The square of the frequency of a transverse optic mode in SrTiO$_3$ as a function of temperature. ($- \cdot - \cdot -$) gives the reciprocal of the static dielectric constant [3.53]. (b) The phonon energy of the same transverse optic mode as shown in (a), measured down to 4 K. (- - -) shows what the phonon energy would be if it continued to vary as $(T - T_c)^{1/2}$ [3.54]

Fig. 3.21b. The sluggish response of ω_{TO} and ε_0^{-1} to T at low temperatures is due to quantum effects (Sect. 3.7 and Problem 3.9.3).

Accurate measurements of the unit cell dimensions of SrTiO$_3$ showed that the crystal becomes tetragonal below a temperature $T_0' = 108$ K. The occurrence of a second-order phase transition is confirmed by the variation of the specific heat, but the dielectric constant and the frequency of the lowest $q = 0$ TO mode apparently vary quite smoothly through this transition. This phase transition is an example of an antiferrodistortive transition devoid of electrical activity as mentioned in Sect. 3.4.3. In this transition alternating rotations of the TiO$_6$ octahedra about a cubic axis produce a doubling of the unit cell in the tetragonal "ordered" phase. The mean rotation angle $\langle \phi \rangle$ represents the order parameter and the vibrational rotations about this mean value represent the soft mode which as a result of the staggered rotations is a mode at the zone boundary of the Brillouin zone. Figure 3.22 shows the frequencies in an acoustic branch of the phonon spectrum of SrTiO$_3$. A low and temperature dependent frequency is found at the zone boundary. Figure 3.23 illustrates the square of this zone boundary mode as a function of temperature [3.55]. The actual change of structure which

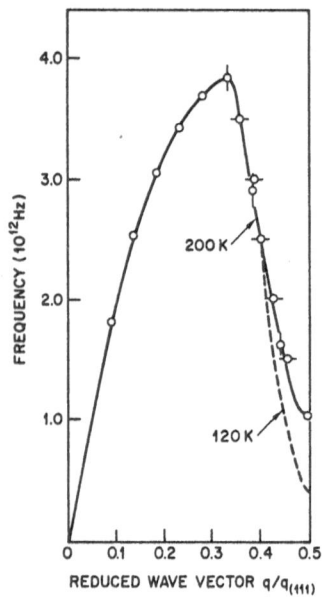

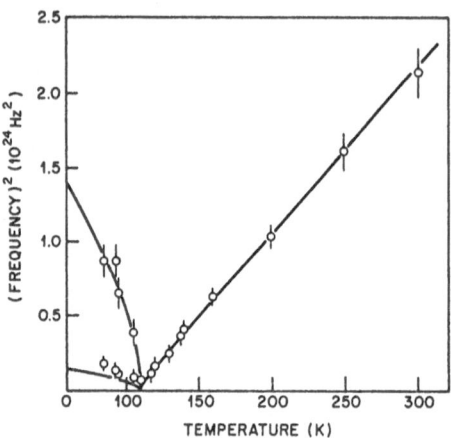

Fig. 3.23. The square of the frequency of the mode of SrTiO$_3$ for which the wavevector q is at the zone boundary in Fig. 3.22, as a function of temperature [3.55]

Fig. 3.22. Frequencies in an acoustic branch of the phonon spectrum of SrTiO$_3$. The wavevector q is parallel to (111); a low and temperature dependent frequency is found at the boundary of the Brillouin zone [3.55]

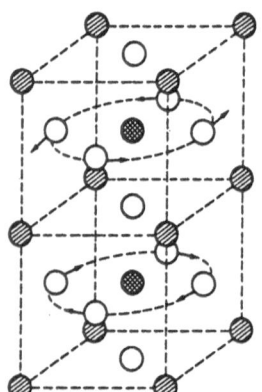

Fig. 3.24. Eigenvector of soft mode of SrTiO$_3$ associated with the zone boundary phonon of Fig. 3.23 [3.31]

occurs at this phase transition is illustrated in Fig. 3.24. One set of oxygen atoms and the Ti and Sr atoms are not displaced. The axis through the undisplaced oxygens defines a unique axis, the z axis in Fig. 3.5. In effect, each regular octahedron which has Ti at its center and O at each of its six corners is rotated about the z axis with the sense of rotation reversed in adjacent unit cells.

An example of a classical displacive system exhibiting a second-order ferroelectric phase transition is tris-sarcosine calcium chloride (TSCC),

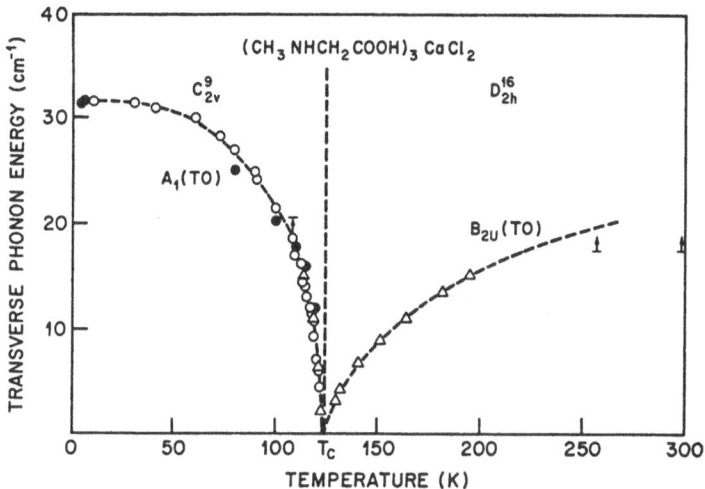

Fig. 3.25. Temperature dependence of the soft mode frequency in TSCC. The triangles are from BWO data, the circles are from Raman measurements [3.56]

$(CH_3NHCH_2COOH)_3CaCl_2$ [3.56–59]. Several recent theoretical approaches have assumed an order-disorder mechanism for the phase transition at $T_c \cong$ 130 K. However, by means of modern far-infrared techniques, in particular by using backward wave oscillator (BWO) spectroscopy in the 3–18 cm^{-1} region revealed an underdamped soft mode both above and below T_c shown in Fig. 3.25 [3.56]. If we use the data of Fig. 3.25 and plot ω_{TO}^2 as a function of temperature the qualitative behaviour near T_c will be the same as that shown in Fig. 3.16b.

TSCC is quite unique among the known ferroelectrics because not only the low-frequency TO mode is soft but also the corresponding LO mode softens considerably [3.56,57]. The soft LO mode is not incompatible with the generalized LST relation (3.39), because the static dielectric constant $\varepsilon_0(T)$ is nearly constant over most of the temperature range; only within a few degrees of T_c does it increase significantly. This unusual behaviour is due to the fact that the Coulomb terms in (3.40a,b) are very small and the similar temperature dependence of ω_{TO} and ω_{LO} arises from the intrinsic instability in the short-range forces, as in a non-ferroelectric displacive phase transition.

3.7 Quantum Ferroelectrics (By U.T. Höchli)

So far we have assumed that the coefficient a in the Landau free energy (3.8) depended linearly on temperature. Quite generally, such linear dependences on temperature are a consequence of anharmonicities treated in terms of classical statistical mechanics [Ref. 3.1, Chap. 5]. In particular, we have renormalized the soft-mode frequency by anharmonic contributions and stated that

$$\omega^2 \cong \omega_0^2 + \text{const}\langle w^2 \rangle_T \ , \tag{3.84}$$

where $\langle w^2 \rangle_T$ was the mean-square displacement of all atoms in the solid. For $T > \hbar\omega/k$, this approach led to write $\omega^2 \sim T - T_c$ justifying the Landau ansatz.

For $T \leq \hbar\omega/k$, however, this approximation breaks down and accordingly a different behaviour is expected for soft-mode frequencies and the LST-related susceptibilities. Deviations from Curie-Weiss behaviour are expected to be strongest near $T = 0$, and since critical effects occur at small $T - T_c$, this implies $T_c \sim 0$. They derive their origin from quantum occupancy of soft modes and this has led to the use of the expression "quantum ferroelectrics" for ferroelectrics whose Curie point falls into the quantum regime. Experimental evidence for quantum ferroelectricity dates back to 1952 when the dielectric susceptibilities of $KTaO_3$ and $SrTiO_3$ where found at variance with the Curie-Weiss expression. Their temperature dependence was expressed in terms of a quantum-mechanical ensemble of anharmonic oscillators [3.60]. Quantum effects were encountered much later in KH_2PO_4 (KDP), a well-studied classical ferroelectric. However, it will behave similarly [3.61] to $KTaO_3$ if it is subject to hydrostatic pressure of about 1.7 GPa, sufficient to reduce T_c to 0 K. This and similar studies showed conclusively that the essential feature of quantum ferroelectrics is the reduction of T_c to 0 K. We first show that it occurs both in order-disorder (KDP) and in displacive-type ($KTaO_3$) ferroelectrics and that the mechanism of reducing T_c, whether pressure or crystal composition, is immaterial for quantum effects.

In Fig. 3.26 we have plotted the inverse susceptibility of KDP vs. temperature on log-log scale [3.62]. The pressure is close to critical, $p_c = 1.7$ GPa, and thus T_c near 0 K in Fig. 3.26b, above critical in Fig. 3.26a and below critical in Fig. 3.26c. Also plotted in these figures is $\varepsilon^{-1} = (T - T_c)^2$, a straight line on log-log scale. Only for the lowest T_c is a fit of ε^{-1} to $(T - T_c)^2$ possible (Fig. 3.26b); at sub- and super-critical pressure (Figs. 3.26a,c) ε^{-1} tends towards the Curie-Weiss expression $T - T_c$. In the immediate neighbourhood of T_c, ε levels off regardless of pressure, as in all ferroelectrics; an effect generally attributed to the presence of impurities and/or inert surfaces and not persued any further here.

We continue the analysis of quantum effects with prototype displacive ferroelectrics. $KTaO_3$ has $T_c < 0$, and hydrostatic pressure reduced it further but partial substitution of Ta by Nb produces a continuous series of ferroelectrics with $0 < T_c < 700$ K [3.3]. The findings reported [3.62] for the mixed crystal $KTa_{1-x}Nb_xO_3$ are similar to those for KDP. When the crys-

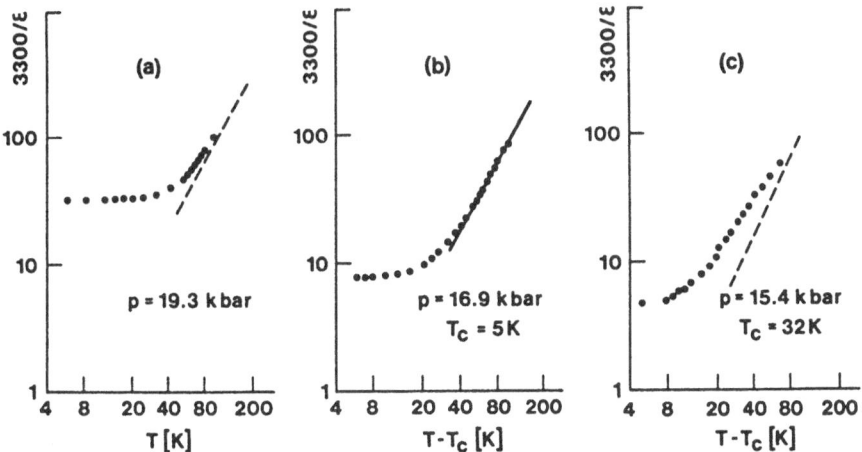

Fig. 3.26a–c. Log-log diagrams of the temperature dependence in the paraelectric phase of the dielectric susceptibility for KH_2PO_4 crystals under hydrostatic pressure. The data sets and the quantum-limit value p_c have been taken from [3.61]. **(a)** $p > p_c$, **(b)** $p \simeq p_c$, and **(c)** $p < p_c$. (- - -) correspond to $\varepsilon = T^{-2}$. The attempted fit is reasonable only at the quantum-limit (case **b** [3.62])

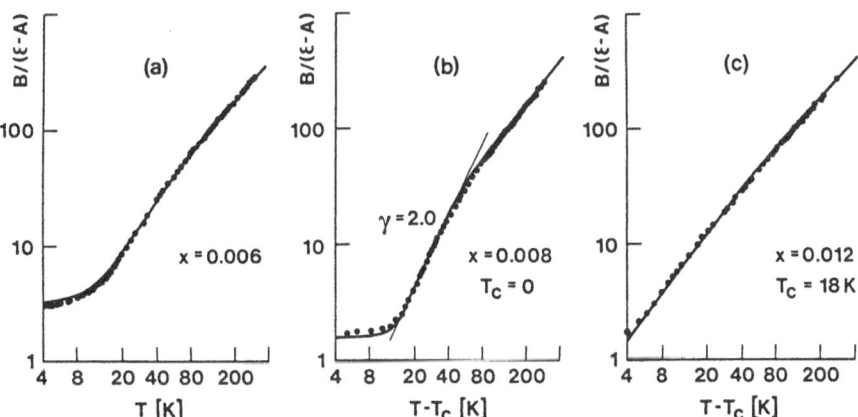

Fig. 3.27a–c. Log-log diagrams of the temperature dependence in the paraelectric phase of the dielectric susceptibility for $KTa_{1-x}Nb_xO_3$ crystals with x near the quantum-limit value x_c : **(a)** $x < x_c$, **(b)** $x = x_c$, and **(c)** $x > x_c$. (——) represent the fits of $B/(\varepsilon - A) = (1/2T_1)\coth(T_1/2T) - T_0$ to our experimental data. In case (b), a fitted power law $B/(\varepsilon - A) = T^\gamma$ with $\gamma \cong 2$ is also shown [3.62]

tal composition is chosen such that $T_c = 0$, (Fig. 3.27b), then there exists a temperature range $10 < T < 50$ K in which $\varepsilon \sim T^{-2}$. This range is absent for neighbouring concentration (Fig. 3.27a,c), but deviations from the Curie-Weiss behaviour are still observed even for $x = 0$ (pure $KTaO_3$).

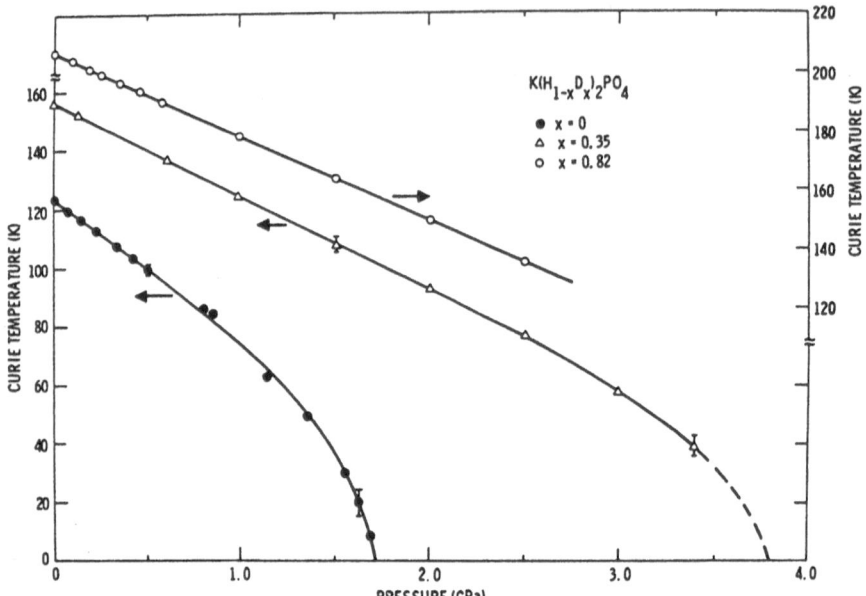

Fig. 3.28. Variation of the transition temperature T_c with pressure for KDP and DKDP (x) samples [3.61]

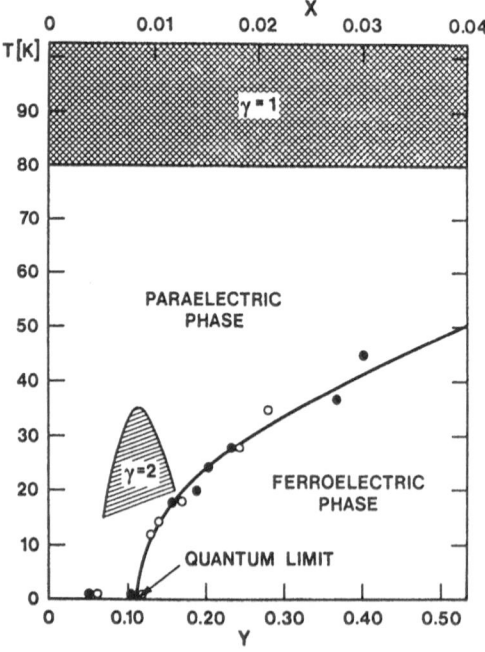

Fig. 3.29. Ferroelectric phase diagram for (•) $KTa_{1-x}Nb_xO_3$ (upper scale) and (0) $K_{1-y}Na_yTaO_3$ (lower scale). (——) is the transition line between paraelectric and ferroelectric states. The crosshatched and the hatched areas indicate the temperature and concentration ranges where $\gamma = 1$ and $\gamma = 2$ hold, respectively [3.62]

We also show in Fig. 3.28 that the transition temperature depends on pressure as $T_c \sim (p_c - p)^{1/2}$ in KDP, where the critical pressure for zero deuteration $p_c = 1.7\,\text{GPa}$. It increases with deuteration. In $\text{KTa}_{1-x}\text{Nb}_x\text{O}_3$ it depends on composition x as $T_c \sim (x - x_c)^{1/2}$ (Fig. 3.29), where $x_c = 0.008$, whereas for Na, $y_c \sim 0.12$. Far away from p_c (or x_c), the phase diagrams are linear, $T_c \sim p_c + \text{const}$ (or $x_c + \text{const}$) as expected for the regime of classical statistics.

There are two major approaches to model phase transitions, the mean-field and the renormalization-group approach and both have been used to compute dielectric susceptibilities in quantum ferroelectrics. In view of their complexity, we quote the results and refer to [3.2a] for derivations and detailed discussions. A simple model is discussed in Problem 3.9.3.

The mean-field approach was originally [3.63] applied to KDP, consisting of PO_4 tetrahedra which are tilted by an angle ϕ with respect to the twofold crystalline axis. The tilts ϕ and $-\phi$ are symmetry-related and give rise to two configurations connected by tunneling. Their symmetrized wave functions thus have a one-to-one relationship with the Ising model of uniaxial spins, $S = \pm 1/2$, and these PO_4 tilts may be expressed in terms of a pseudo-spin Hamiltonian

$$H^{\text{eff}} = -\Delta S_l^* - \sum_{ll'} J_{ll'} S_l^z S_{l'}^z \;, \tag{3.85}$$

where Δ is the tunnel splitting and $J_{ll'}$ the interaction between states in adjacent unit cells. For the case that the actual field at site l, $h_l = \sum_{l' \neq l} J_{ll'} S_{l'}$ can be replaced by the mean field

$$h_l = \sum_{l' \neq l} J_{ll'} \langle S_{l'} \rangle \;, \tag{3.86}$$

the result is

$$\langle S^z \rangle = \tfrac{1}{2} \tanh \Delta / 2kT \tag{3.87}$$

and the concomitant susceptibility is

$$\varepsilon = (2\Delta \coth \Delta / kT - \sum J_{ll'})^{-1} \;. \tag{3.88}$$

For small Δ / kT (3.88) reduces to the Curie-Weiss expression

$$\varepsilon = (kT - \sum J_{ll'})^{-1} \;. \tag{3.89}$$

Equations (3.88 and 89) are seen to fit the data of Figs. 3.26, 27 reasonably well. We note that an equation of the type (3.88) also holds for an ensemble of uncoupled anharmonic quantum oscillators [3.60]. Setting $2\Delta = \sum J_{ll'}$ in Eq. (3.88) leads to $\varepsilon = \infty$ at $T = 0$. At this point both $\varepsilon(T)$ and $T_c(\sum J_{ll'})$ are highly singular and inconsistent with the data. Quite generally, replacement of the actual field by its mean value leads to inconsistencies near T_c.

In this region a renormalization-group (RG) treatment of a Hamiltonian of the form (3.85) leads to [3.64],

$$\varepsilon \sim T^{-2} \quad \text{and to} \tag{3.90}$$

$$T_c \sim (p - p_c)^{1/2} \ , \tag{3.91}$$

where $p = \sum_{ij} J_{ij}$, and p_c is a critical set of interactions $\sum_{ij} J_{ij}^c$ setting $T_c = 0$. The expression (3.90) fits the experimental results with better accuracy than (3.88) and exactly under those conditions for which RG quantum theory applies, namely $T_c \sim 0\,\mathrm{K}$.

The dependence of T_c on $\sum J_{ij}$ can be verified only indirectly. Since J_{ij} is a measure for attraction between cells, it should vary with distance and chemical composition provided this leaves the structure invariant. Identifying $J_{ij} + \text{const}$ as pressure leads to $T \sim (p - p_c)^{1/2}$, a result which describes nicely the data in KDP (Fig. 3.28). Identifying J_{ij}' as Nb content x in $KTa_{1-x}Nb_xO_3$, we find $T_c \sim (x - x_c)^{1/2}$ also confirming the data in this compound (Fig. 3.29).

Quantum ferroelectricity has so far established in two classes of crystals: the perovskites (represented by $KTaO_3$) whose T_c can be made $0\,\mathrm{K}$ by adding impurities, and KDP where $T_c = 0$ at a pressure of $1.7\,\mathrm{GPa}$. To all appearances, the behaviour of the quantum limit is identical in these two crystals. Investigations on pressure-reduced quantum effects are not as numerous as those in mixed crystals. This is probably due to the difficulty of performing supporting experiments, like phonon spectroscopy, in pressure gauges. On the other hand, in mixed crystals, interactions have a random component on a local scale and this random component may modify quantum-ferroelectric local properties [3.65].

3.8 Disordered Polar Systems (By U.T. Höchli)

The lattice-dynamical approach showed that interactions between different cells were capable of establishing order below a certain temperature. In this approach, all interactions between adjacent cells were considered identical and accordingly polar order resulted in infinite correlation.

In a single crystal, some lattice sites may be replaced by impurities containing a dipole moment. The resulting dipolar interactions

$$J_{ij} = (\boldsymbol{p}_i \cdot \boldsymbol{p}_j)/r_{ij}^3 - 3(\boldsymbol{p}_i \cdot \boldsymbol{r}_{ij})(\boldsymbol{p}_j \cdot \boldsymbol{r}_{ij})/r_{ij}^5 \tag{3.92}$$

have both positive and negative contributions. If the occupations of impurities is random, J_{ij} has equal probability of being positive and negative. If these interactions are sufficiently strong, as we expect them to be for large dipole concentration and/or in a polarizable lattice, then they may force the dipoles into a polar phase. By this we mean that they assume fixed directions in space, but in view of their randomness not necessarily along a single polar axis.

Examples of this kind are OH ions replacing Cl in alkali halides [3.66] and Li replacing K (at an off-center position due to the large misfit in ionic size) in KTaO$_3$ [3.67]. Of the many investigations made in these disordered, polar crystals, we emphasize the results derived from an experiment involving phonons.

The propagation properties of phonons generally depend on the direction of the polar axis in a crystal. In particular, the transverse optical mode which is triply degenerate in the cubic phase, splits into a singlet and a doublet in a polar crystal. This splitting has been observed by Raman-scattering experiments in many polar crystals and, in particular [3.68] in KTaO$_3$:Li. In Fig. 3.30b,c the two Raman lines are shown for a crystal which was cooled while an electric field was applied. The polarization characteristics of these two lines allow the identification of their symmetry, A and E, consistent with C$_{4v}$ symmetry of the polar phase and consistent with uniform alignment of the Li dipoles by the field. If instead K$_{1-x}$Li$_x$TaO$_3$ is cooled at zero field, then a broad isotropic line will occur in the Raman-scattering spectrum (Fig. 3.30a). Such spectra can be obtained if it is assumed that k-vector conservation is violated in the Raman process [3.69]. In zero-field cooled samples the Raman susceptibility and the concomitant Li displacements are thus uncorrelated on the scale of a phonon wavelength. Such small correlation lengths ("polar clusters") at and below T_c are at variance with the domain structure found in nonrandom ferroelectrics.

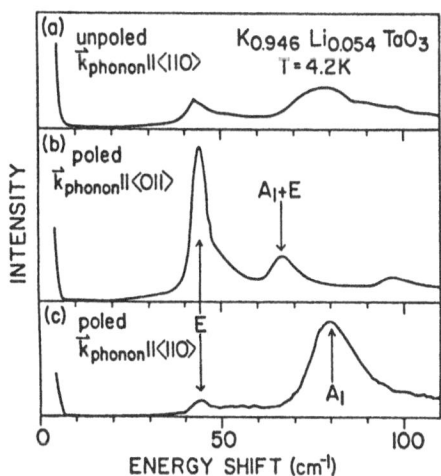

Fig. 3.30. Raman spectra of KTaO$_3$ containing 5.4 mol % Li at 4.2 K: **(a)** the unpoled sample, **(b)** and **(c)** the poled samples using two different scattering geometries. Polarizations included are $Y(ZZ)X$ and $Y(ZY)X$, where X, Y, Z are pseudocubic axes [3.68]

69

As a consequence of finite-size correlation, the susceptibility must remain finite and a considerable dispersion is expected in the temperature range where these regions of finite correlation, called clusters, form: The larger they grow, the slower they become; ε should thus be larger for low ω and go through a maximum at a frequency-dependent temperature. Such findings have been reported [3.70] in Fig. 3.31 and we note that analogous effects exist for magnetic analogs [3.71] like Cu:Mn, where the spin-carrying Mn ion plays the role of the dipole. In both systems, the responses are isotropic at all temperatures, if zero-field cooled, and complex hysteresis and long-time effects are observed. Random electric and magnetic systems share their isotropic responses and the relaxational character of the transition with glasses and for that reason often carry that name, glass. We note here that interpretation of experimental results in terms of finite-size correlations have their pitfalls and that a combination of results from several experiments is required to obtain clarity [3.72]. In these the phonons play a leading role since they probe lattice distortions at the proper wavelengths. Theories trying to describe [3.68] electric-dipole glasses and their magnetic analogs have faced severe problems. The simplest approach [3.73] was an attempt to solve the Hamiltonian (in spin variables)

$$ H = \sum J_{ij} S_i S_j + \sum h S_i \; , \tag{3.93} $$

where the random interaction is taken independent of the pair i, j and has a Gaussian distribution with variance J and mean value $\bar{J}$. The mean-field solution of (3.93) predicts a susceptibility maximum

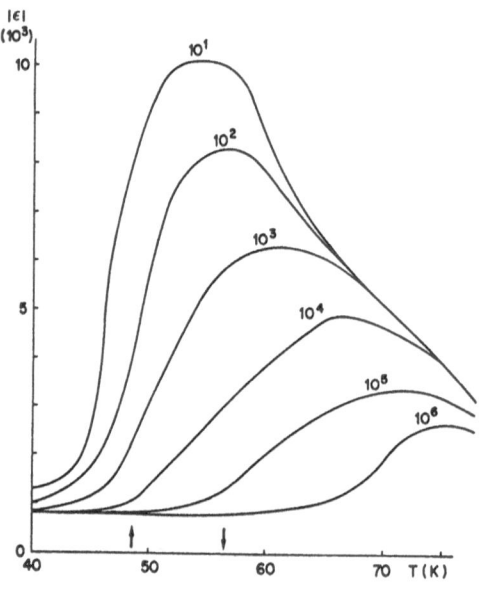

Fig. 3.31. Dielectric susceptibilities of KTaO$_3$:Li as a function of temperature. Labels stand for the measured frequencies (cps), arrows for the maximum of the dielectric dispersion step (49 K) and the stability limit of remanent polarizaton (56 K). Li concentration is 0.026 [3.70]

$$\partial S / \partial h = (T - T_c)^{-1} , \tag{3.94}$$

where $T_c = J/k$. Below T_c (3.93) produces an infinite number of non-symmetry-related states defying thermal averaging. Algebraic decay of the polarization $(P \sim t^{-v})$ is predicted on the basis of a relaxational approach within mean-field theory [3.74]. Numerical simulations have confirmed the general features of glassy behaviour, however, they are limited to small sample size and rely on extrapolation to finite size. A particularly simple case is a closed one-dimensional ring consisting of N particles [3.75]. If their interaction is random, then the ring will show characteristic properties seen in experimental glasses for N as low as seven. There is a large amount of literature on this subject, mostly in magnetic language, but since the interaction between "spins" is taken by just probabilities for plus and minus they also include the dipolar case. At the time of writing, disordered systems are an active field of research in which phonon spectroscopy plays a major role, but where many questions are left open.

3.9 Problems

3.9.1 Temperature Dependence of the Soft Mode in Displacive Phase Transitions

a) Show that in a first-order phase transition

$$\mu \omega_{TO}^2 (T \geq T_c) = \frac{\alpha}{g^2 v_a} (T - T_0) ,$$

$$\mu \omega_{TO}^2 (T \leq T_c) = \frac{1}{g^2 v_a} [3b P_s^2(T) + 5c P_s^4(T) - \alpha(T_0 - T)] ,$$

where $P_s^2(T)$ is given by (3.16b). From these results derive the relations (3.62,67).

b) Show that in a second-order phase transition

$$\mu \omega_{TO}^2 (T \geq T_c) = S^* \gamma (T - T_c) ,$$

$$\mu \omega_{TO}^2 (T \leq T_c) = 2 S^* \gamma (T_c - T) .$$

Hint: Use (3.60 with $P \cong w/g v_a$ and evaluate $(d^2 \Phi / \partial w^2)_0$ and $(\partial^2 \Phi / \partial w^2)_{w_s}$, where $\Phi \cong v_a (F - F_0)$ and $w_s = g v_a P_s$. To derive $\mu \omega_{TO}^2 (T \leq T_c)$ for a second-order phase transition use (3.30).

3.9.2 Polarizability Catastrophe for a System of Two Neutral Atoms

Consider a system of two neutral and identical atoms separated by a distance R, each having an electronic polarizability α; the external field E is applied parallel to the direction of $\boldsymbol{R}$. Find the relation between R and α for such a system to be "ferroelectric" or polar.

Result:

$$R^3 = 2\alpha \ .$$

Hint: Consider [Ref. 3.78, Problem 3.4.1]. The electronic dipole moment is $p_{\parallel}(R) = \alpha_{\parallel}(R)E$; $\alpha_{\parallel}(R)$ becomes infinite for $R^3 = 2\alpha$.

Compare this result with the condition (3.45) in the case $\alpha_i = 0$, i.e., $z^* = 0$ corresponding to a simple cubic crystal composed of neutral and identical atoms; in the rocksalt structure $v_a = a^3/4$ with $a = 2R$, leading to $\alpha_e \cong (3/2\pi)R^3 \sim R^3/2$.

3.9.3 Classical and Quantum Ferroelectrics

If the transition temperature T_c of a ferroelectric crystal is sufficiently high, classical theory applies and $\varepsilon(T)$ follows the Curie-Weiss law. If, however, T_c is close to $T = 0$, quantum effects become important and deviations from the Curie-Weiss law are expected (Sect. 3.7). To demonstrate this situation consider the incipient ferroelectric $SrTiO_3$ (Fig. 3.5). The lowest $q = 0$ TO mode is a quasi-soft mode as shown in Fig. 3.21; in this mode the Ti and Sr ions displace along a (100) direction while the oxygen ions displace in the opposite direction. Each different set of ions displaces by different amounts, but the general situation is a relative motion of the positive and negative ions. To simplify the model we disregard the motion of the Sr ions, that is, we consider only the relative motion of the Ti ions against the oxygen ions; it can be shown that despite this approximation we are correctly treating the major part of the ion motion [3.76]. For the total potential energy we write

$$\Phi(w) = \Phi^{(s)}(w) + \Phi^{(c)}(w) + \Phi^{\text{ext}}(w) \ . \tag{3.9.1}$$

Here w is the relative Ti-O displacement, $\Phi^{(s)}(w)$ is the short range (Ti-O) interaction energy, $\Phi^{(c)}(w)$ is the Coulomb energy resulting from the interactions between the displacement dipoles [Ref. 3.1, Sect. 4.2], and $\Phi^{\text{ext}}(w)$ is the potential energy due to the external field $E = E_0 \exp(i\omega t)$. For $\Phi^{(s)}(w)$ we write

$$\Phi^{(s)}(w) = \tfrac{1}{2}S^* w^2 + \tfrac{1}{4}Aw^4 \ , \tag{3.9.2}$$

where in addition to the harmonic term we have included a fourth-power anharmonic term. For a TO mode at $q = 0$ the Coulomb energy $\Phi^{(c)}(w)$ is negative [Ref. 3.1, Sect. 4.2] and accordingly we write

$$\Phi^{(c)}(w) = -\tfrac{1}{2}C^* w^2 \ , \tag{3.9.3}$$

where C^* is a positive force constant. The interaction energy with the external field is given by

$$\Phi^{\text{ext}}(w) = -z^* e\, E w \; , \tag{3.9.4}$$

where $z^* e$ is an effective charge. The total potential energy is then

$$\Phi(w) = \tfrac{1}{2}(S^* - C^*)w^2 + \tfrac{1}{4}A w^4 - z^* e\, E w \; . \tag{3.9.5}$$

The equation of motion is obtained from $m\ddot{w} = -\partial\Phi/\partial w$, where m is an effective mass, thus

$$m\ddot{w} + (S^* - C^*)w + A w^3 = z^* e\, E_0\, \mathrm{e}^{\mathrm{i}\omega t} \; . \tag{3.9.6}$$

In the following we assume that the dipole-dipole interaction is strong enough to lead to a destabilization of the harmonic system at $w = 0$, i.e., we assume $C^* > S^*$ and $R = C^* - S^* > 0$. The system is then stabilized by the anharmonic term $A w^3$ in (3.9.6). In order to linearize the equation of motion we replace w^3 by $2\langle w^2\rangle_T w$, where $\langle w^2\rangle_T$ is the thermal average. We have thus introduced an effective quadratic potential energy which depends on T, and the equation of motion can be written in the form

$$m\ddot{w} + (3A\langle w^2\rangle_T - R)w = z^* e\, E_0\, \mathrm{e}^{\mathrm{i}\omega t} \; . \tag{3.9.7}$$

a) Classical Ferroelectrics

At high temperatures $\langle w^2\rangle_T$ is given by

$$\langle w^2\rangle_T \cong k_{\mathrm{B}}T/m\omega_0^2 \; , \tag{3.9.8}$$

where $\omega_0 = (S^*/m)^{1/2}$ is the harmonic frequency due to short-range forces.

Show that the susceptibility χ defined by $P = (z^* e/v_a)w = \chi E$ (P : polarization, v_a : volume of the unit cell) is given by

$$\chi = \frac{(z^* e)^2/v_a m}{\omega_s^2 - \omega^2} \; , \tag{3.9.9}$$

where the soft-mode frequency is given by

$$\omega_s^2 = \frac{B}{m}(T - T_{\mathrm{c}}) \; , \quad \text{where} \tag{3.9.10}$$

$$B = 3A k_{\mathrm{B}}/S^* \quad \text{and} \tag{3.9.11}$$

$$T_{\mathrm{c}} = \frac{R}{B} = \frac{1}{B}(C^* - S^*) \; . \tag{3.9.12}$$

The static suceptibility is obtained from (3.9.9) as

$$\chi(\omega = 0) = \frac{C}{T - T_{\mathrm{c}}} \; , \tag{3.9.13}$$

where $C = (z^* e)^2/B v_a$. This is the Curie-Weiss law.

b) Quantum Ferroelectrics

If T_c is close to $T = 0$ the zero-point motion becomes important and instead of (3.9.8) we must write

$$\langle w^2 \rangle_T \cong \frac{\hbar}{2m\omega_0} \coth(\hbar\omega_0/2k_B T) \ . \tag{3.9.14}$$

Show that for quantum ferroelectrics our simple model predicts

$$\omega_s^2 = \frac{B}{m}\left[\frac{T_1}{2}\coth\left(\frac{T_1}{2T}\right) - T_c\right] \quad \text{and} \tag{3.9.15}$$

$$\chi(\omega = 0) = \frac{C}{(T_1/2)\coth(T_1/2T) - T_c} \ , \tag{3.9.16}$$

where $T_1 = \hbar\omega_0/k_B$ represents the dividing temperature between the quantum mechanical and classical region. Note that at high temperatures (3.9.16) becomes identical with (3.9.13). Make figures of $\omega_s^2(T)$ and $\chi(\omega = 0, T)$ for $T_c = 0, \pm T_1/4$.

In spite of this apparent ability to provide a reasonable fit to the observed susceptibilities of $SrTiO_3$ and $KTaO_3$, Eq. (3.9.16) is in contradiction with the limiting behaviour of $\chi(T \to 0)$ as predicted by the more recent quantum theoretical treatments of incipient ferroelectrics with $T_c = 0$ [3.64, 77]. The variation of χ with temperature predicted by these more recent theories is discussed in Sect. 3.7.

3.9.4 Polarizability Model

Consider a vibrating atom consisting of a core with mass m, displacement u, and a shell with mass $m_e \equiv 0$ and displacement v in contact with a heat bath at temperature T. The attractive Coulomb interaction between core and shell is described by the harmonic force constant $g < 0$, while the repulsive core-shell interaction is modeled by the quartic anharmonic force-constant $h > 0$ (Fig. 3.32). In this model the relative core-shell motion is that of a particle in a double-well potential. Show that the frequency of the "soft mode" is given by

$$\omega_f^2 = \omega_0^2 \frac{g(T)}{2f + g(T)} \ ,$$

where $g(T) = g + g_T = g + 3h\langle w^2\rangle_T$, $w = v - u$, and $\omega_0^2 = 2f/m$. This result is similar to (3.78, 79).

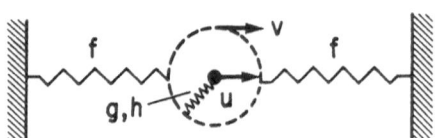

Fig. 3.32. Polarizability model

Hint: Start from the equations of motion of core and shell and linearize these equations by using the approximation $hw^3 \cong 3h\langle w^2\rangle_T w = g_T w$. Eliminate v from the adiabatic condition to obtain an effective harmonic equation of motion for u.

4. Thermal Conductivity

In this chapter we consider experimental and theoretical aspects of thermal conductivity in insulators. Section 4.2 contains a discussion of the experimental determination of thermal conductivity. The elementary kinetic theory of thermal conductivity is presented in Sect. 4.3, while Sect. 4.4 is devoted to the formal theory of thermal conductivity which is based on the Boltzman transport equation and a phenomenological ansatz for the collision term (relaxation time approach). In Sect. 4.5 a discussion is given for the relaxation times of insulators: boundary scattering, defect and impurity scattering as well as phonon-phonon scattering are considered. In the present context phonon-phonon scattering is of special importance. Section 4.5.4 therefore contains a qualitative discussion of the microscopic approach, the role of normal and umklapp processes, and the relaxation times for umklapp processes. Sections 4.6 and 7 contain a short discussion of the thermal conductivity of glasses and of metals and alloys. A discussion of second sound is given in Sect. 4.8, while Sect. 4.9 contains a number of illustrating problems.

4.1 General Remarks

The heat transport by lattice waves in solids is governed by anharmonicity [Ref. 4.1, Chap. 5], by the various imperfections of the crystal lattice, and by external boundaries. Moreover, in the case of metals and semi-metallic solids thermal conductivity is also governed by the conduction electrons. Not only may many different factors influence the thermal conductivity, but the processes which provide the principal sources of thermal resistance may vary in different temperature regions in any one material.

The thermal conductivity σ_t of a solid is most easily defined with respect to the steady-state flow of heat down a long rod with a temperature gradient $\partial T / \partial x$:

$$J = -\sigma_t \frac{\partial T}{\partial x} \ . \tag{4.1}$$

J is the flux of thermal energy (energy transmitted across unit area per unit time); accordingly σ_t is often expressed in units of cal/cm $\cdot$ s $\cdot$ deg or W/cm $\cdot$ deg. σ_t is positive, since the thermal current flows opposite to the direction of the temperature gradient.

During the process of heat transport, the energy does not simply enter one end of the specimen and proceed directly in a straight path to the other end; the energy rather diffuses through the specimen, suffering many collisions. If the energy were propagated directly through the specimen without deflection, then the expression for the thermal flux would not depend on the temperature gradient, but only on the difference in temperature, ΔT, between the ends of the specimen, regardless of the length of the specimen. It is the random nature of the heat transport process that brings the temperature gradient into the expression for the thermal flux.

In the following we shall confine ourselves to a discussion of thermal conductivity of insulators. This is a difficult subject which we can only treat in a rather superficial way; nevertheless the principal features of thermal conduction in insulating solids may be understood from essentially qualitative arguments. An approximate theory of the effect of anharmonic interactions on thermal conductivity has been given by *Debye* [4.2a], and *Peierls* [4.2b] has considered the problem in great detail. More recent studies of thermal conductivity have been performed by *Herring* [4.3], *Callaway* [4.4], *Nettleton* [4.5], *Holland* [4.6], *Klemens* [4.7], *Erdös* [4.8, 39, 40]; for a review see *Carruthers* [4.9].

In a perfectly harmonic and defect-free insulating crystal the phonons which are the carriers of energy transport are not coupled and propagate as free particles. Therefore, if a distribution of phonons is established that carries a thermal current (as a simple consequence of statistical fluctuations, leading, for example, to an excess of phonons with similarly directed group velocities), this distribution will remain unaltered in the course of time and the thermal current will remain forever undegraded. One would therefore have an energy transport without a thermal gradient, and hence, according to (4.1), an infinite thermal conductivity.[1]

We can also consider the mean-free path of the phonons. In a defect free and infinitely large crystal the mean-free path l is the mean distance between successive collisions of phonons. In the harmonic approximation there are no phonon-phonon collisions and therefore l is infinite. In an anharmonic crystal, however, l and σ are limited by phonon-phonon collisions. For this reason anharmonicity plays a central role in thermal conductivity.

We have already mentioned that in a real crystal not only phonon-phonon collisions, but also collisions of phonons with defects, impurities and the boundaries have to be considered. The mean free path l depends therefore on all these interactions, and it is not surprising that there is no "exact" theory of thermal conductivity, but only approximate theories which apply in particular ranges of temperature.

---

[1] The above argument leading to an infinite thermal conductivity is frequently encountered in the literature. It should be mentioned, however, that for an asymmetric (non-equilibrium) distribution one cannot define precisely a temperature T, which is, after all defined by the Bose-Einstein distribution.

4.2 Experimental Determination of Thermal Conductivity

The experimental determination of the thermal conductivity is based either on the steady-state heat flow method or on the heat pulse transmission method.

In the steady-state heat-flow method the heat H is supplied at one end of the crystal by a suitable heater and extracted at the other end of the crystal by a large copper plate which acts as a heat sink (Fig. 4.1). The temperature T of the copper plate at the base of the crystal is adjusted with respect to the helium bath by means of an auxiliary heater and a fixed thermal leakage resistance. Two carbon resistance thermometers are placed along the length of the crystal in order to measure the average temperature and the temperature gradient. The crystal, heaters, thermometers, and heat sink are all placed inside an evacuated copper can which is submerged in the liquid helium bath. An experimental arrangement which has been used to measure the low temperature thermal conductivity of Lif [4.10] and KCl [4.11] is shown in Fig. 4.1.

The thermal conductivity $\sigma_t(T)$ is determined by measurement the heat input H to the crystal, the temperature difference ΔT between the

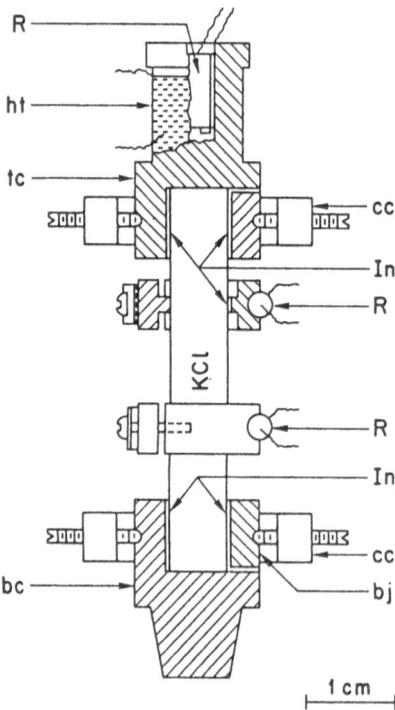

Fig. 4.1. Arrangement for experimental determination of the thermal conductivity of a KCl crystal with heater (*ht*), clamps (*cc*), and thermometers (*R*); In: indium foil; tc: top cap, bc: bottom cap [4.11] (*see text*)

carbon thermometers, the distance b between the thermometers, and the cross-sectional area A of the crystal:

$$\sigma_t(T) = -\frac{Hb}{A \Delta T} \ .$$

(4.2)

Comparing with (4.1) we have $\partial T/\partial x = \Delta T/b$ and $J = H/A$. In (4.2) T is the arithmetic average of the temperatures of the two resistance thermometers; ΔT is usually of the order of 0.01 to 1 K and H is of the order of 10 to 200 mW [4.11]. An experimental arrangement which has been used to measure the thermal conductivity of polydiacetylene single crystals by the static method in the temperature range between 50 mK–50 K has been described by *Wybourne* et al. [4.46–48].

The experimental study of the propagation of heat pulses in solids is not only of importance for measuring thermal conductivity but also for studying the propagation of second sound (Sect. 4.8). The combination of spatial and temporal resolution of the detected signals has made it possible to obtain information not available from standard steady-state thermal conductivity measurements. The heat-pulse experiments in solids require as basic elements a small heater or thermal transducer to produce an excitation of known pulse width at a given time, and a thermal receiver whose response is proportional to the incident thermal flux. The transducer and receiver are evaporated as thin films on opposite polished surfaces of the crystal. This assures a very good interfacial contact and a fast thermal detector, the latter mainly due to the small heat capacity of the film. Heat pulses can be produced by short-pulse current generators, by microwave generators, or by pulsed lasers. The first method utilizes direct joule heating by a pulsed current passing through a small thin-film resistor made by evaporation. The second method utilizes the power from a pulsed microwave source which is absorbed in a thin metal film evaporated onto one surface of a dielectric crystal. Third, the light absorbed in a metallic layer, typically from a gallium arsenide or a giant-pulse ruby laser can be used as an effective heat-pulse source. Figure 4.2 shows the experimental apparatus used for the investigation of heat pulses in sapphire as a function of temperature [4.12]. An evaporated In-Sb alloy film (~2000 Å thick) over the entire front surface of the crystal was used to absorb the optical radiation from a giant-pulse ruby laser acting as the source of heat pulses; a pure-indium thin-film bolometer shown in Fig. 4.2b was used as a detector.

The quantity which is usually measured in the pulse method is the thermal diffusivity α_t, rather than the thermal conductivity σ_t [4.13]. The relation between the two quantities is

$$\sigma_t = \varrho c_p \alpha_t \ ,$$

(4.3)

where $c_p(T)$ is the specific heat at constant pressure, and ϱ is the density.

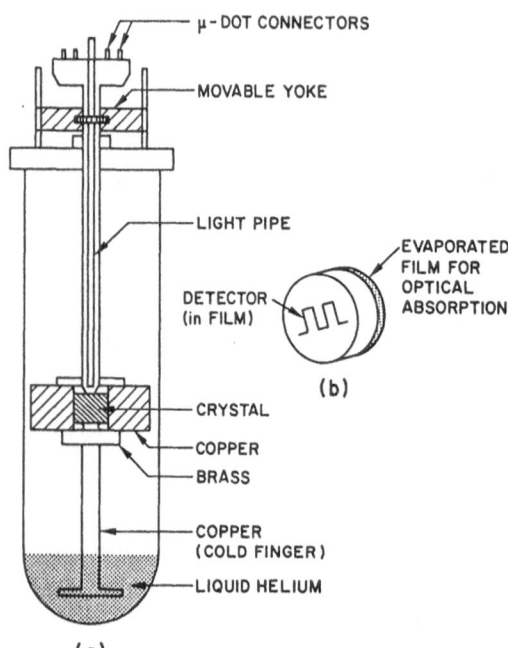

μ-DOT CONNECTORS

MOVABLE YOKE

LIGHT PIPE

DETECTOR (in FILM)

EVAPORATED FILM FOR OPTICAL ABSORPTION

CRYSTAL

COPPER

BRASS

COPPER (COLD FINGER)

LIQUID HELIUM

(b)

(a)

Fig. 4.2. (a) Apparatus for measuring heat pulses as a function of temperature. The helium level with respect to the crystal is adjusted by means of the moveable yoke which determines the ambient temperature of the crystal. (b) Sapphire sample with heater and detector films [4.12]

Because of the simplicity of this equation, it is possible to determine thermal conductivity from the measurement of thermal diffusivity, assuming that the quantities c_p and ϱ are known, can be measured separately, or can be estimated with sufficient accuracy. It can be shown that α_t is given by the expression

$$\alpha_t = \frac{1.37 L^2}{\pi^2 t_{1/2}} \ , \tag{4.4}$$

where L is the thickness of the sample and $t_{1/2}$ is the time at which the back surface temperature reaches half of its maximum value [4.13].

In addition to the determination of thermal conductivity, considerable direct information on phonon-scattering processes is available from heat pulse experiments as a function of temperature, crystal size, and the degree of polish of the boundaries. The shape of the heat pulse is an indicator of phonon scattering, since the heat that arrives at the detector at times other than that corresponding to the LA and TA group velocities must have arrived by an indirect trajectory. Even the heat that arrives in the sharp part of the pulses is composed, in part, of phonons that have arrived with some degree of scattering, since these detected pulses are broader than the input pulse.

4.3 Lattice Thermal Conductivity: Elementary Kinetic Theory

The concept of the phonon gas [Ref. 4.1, Sect. 2.2.4] suggests to consider the thermal conductivity of a gas of particles which are subjected to mutual collisions. We suppose that between two collisions the interactions of a particle with other particles are negligibly small, that is, the particles behave as free particles between two successive collisions. We further assume that a particle will collide with an other particle with a probability dt/τ in the time element dt. Consequently, in statistical average a particle moves as a free particle during the time τ; τ is the mean time between two collisions and is called the *relaxation time*. The frequency $\omega_c = \tau^{-1}$ is the *collision frequency*, and the *mean free path* is given by

$$l = v\tau , \tag{4.5}$$

where v is the mean velocity of the particles. It is assumed that local thermal equilibrium is achieved only through collisions, that is the particles thermalize through collisions. If, for example, a particle leaves a region 1 with local temperature T_1, its mean velocity $v_1(T_1)$ is appropriate to T_1. If this particle arrives at a region 2 with local temperature $T_2 < T_1$, the velocity of the incoming particle adjusts to the mean velocity $v_2(T_2) < v_1(T_1)$ by a single collision. Suppose a small temperature gradient is imposed along the x-direction in the gas of particles. Each particle at a given point will contribute to the thermal current density in the x-direction an amount equal to the product of the x-component of its velocity with its contribution to the energy density. The average contribution of a particle to the energy density depends on the position of its last collision. The thermal flux at the position x is carried by particles whose last collision was, on the average, a distance $l = v\tau$ away from x (Fig. 4.3). Particles with velocities making an angle θ with the x-axis at x collided last at a point P a distance

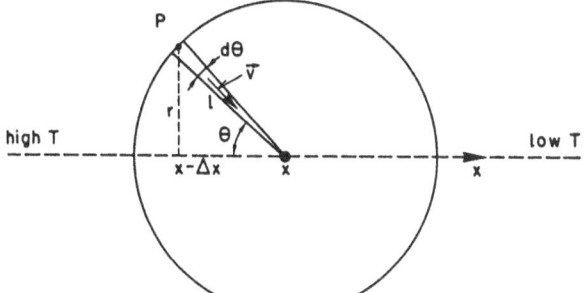

Fig. 4.3. Heat propagation by phonons in the presence of a uniform temperature gradient along the x axis (*see text*)

81

$\Delta x = l\cos\theta$ up the temperature gradient, and therefore carry an energy density $E(x - \Delta x) = E(x - l\cos\theta)$ with velocity $v_x = v\cos\theta$. Let n be the total number of particles and $n(\theta)$ the number of particles with velocity $v_x = v\cos\theta$. From $n(\theta) = n\, dF(\theta)/4\pi l^2$ and $dF(\theta) = 2\pi r l\, d\theta = 2\pi l^2 \sin\theta\, d\theta$ (Fig. 4.3), we have $n(\theta) = (n/2)\sin\theta\, d\theta$, and the thermal flux of these particles is

$$J(\theta) = n(\theta)v_x E(x - \Delta x) = \tfrac{1}{2}nv\, E(x - \Delta x)\sin\theta\,\cos\theta\, d\theta \ .$$

Developing $E(x - \Delta x) \cong E(x) - (\partial E/\partial x)\Delta x$ gives

$$J(\theta) = \frac{1}{2}nv\, E(x)\sin\theta\,\cos\theta\, d\theta - \frac{1}{2}nvl\frac{\partial E}{\partial x}\sin\theta\,\cos^2\theta\, d\theta \ . \tag{4.6}$$

The net thermal flux is obtained by integrating over θ. The integral of the first term of (4.6) vanishes while the value of the second integral is $2/3$, and we obtain

$$J = -\frac{1}{3}nvl\frac{\partial E}{\partial x} = -\frac{1}{3}nvl\frac{\partial E}{\partial T}\frac{\partial T}{\partial x} = -\frac{1}{3}vlC\frac{\partial T}{\partial x} \ , \tag{4.7}$$

where $C = n\,\partial E/\partial T$ is the specific heat. Comparing (4.7) with (4.1), we obtain for the thermal conductivity

$$\sigma_t = \tfrac{1}{3}vlC = \tfrac{1}{3}v^2\tau C \ . \tag{4.8}$$

This is the result obtained from the kinetic theory of gases.

The relation (4.8) can be applied to the heat transport of an insulator whose thermal properties are approximated by the Debye model [Ref. 4.1, Sect. 3.5]. $C(T)$ is the specific heat and the velocity v is temperature independent in a first approximation. (This is in contrast to a classical gas where $v^2 \sim k_B T$). The crucial quantity is the phonon collision rate τ^{-1}; the question of the temperature dependence of τ^{-1} is one of great subtlety and complexity, which took many years to understand fully.

In a crystal with more complex phonon dispersion σ_t must be a sum of terms like Eq. (4.8) over all phonons $s = (qj)$:

$$\sigma_t = \frac{1}{3}\sum_s v_s l_s C_s = \frac{1}{3}\sum_s v_s^2 \tau_s C_s \ . \tag{4.9}$$

Here C_s is the contribution of the mode s to the specific heat, v_s is the group velocity and τ_s is the relaxation time of the phonon s. In the following the relation (4.9) will be derived on the basis of the Boltzmann transport equation.

4.4 Formal Theory of Thermal Conductivity

In order to calculate the heat current density it is best to proceed from the particle point of view, i.e. from phonons which are equivalent to the system of lattice vibrations. The phonons carry a heat current which is the sum of the heat currents carried by all normal modes $s = (qj)$:

$$J = \sum_s E_s v_s , \qquad (4.10)$$

where $E_s = \hbar \omega_s (n_s + 1/2)$ is the energy [Ref. 4.1, Sect. 2.2.4] and v_s the group velocity of the phonons [Ref. 4.1, Sect. 3.5.1]; n_s is the number of phonons in state s (per unit volume). Since the modes are distributed symmetrically in q-space, that is, to every mode $s = (qj)$ there corresponds a mode $-s = (-qj)$ with $\omega_{-s} = \omega_s$ and $v_{-s} = -v_s$, the zero point energy carries no heat current, and we obtain

$$J = \sum_s n_s \hbar \omega_s v_s . \qquad (4.11)$$

Under equilibrium conditions, with a uniform temperature T, the average value of n_s is the mean occupation number given by [Ref. 4.1, Sect. 2.2.4]

$$\bar{n}_s = \bar{n}_{-s} = [\exp(\hbar \omega_s / k_B T) - 1]^{-1} , \qquad (4.12)$$

and hence in thermal equilibrium, with a uniform temperature, the heat current J_{eq} vanishes:

$$J_{eq} = \sum_s \bar{n}_s \hbar \omega_s v_s = 0 . \qquad (4.13)$$

If one allows for interactions between the phonons (anharmonic crystal), energy is interchanged slowly between the lattice waves and in the absence of a temperature gradient the occupation numbers n_s tend to the equilibrium distribution, and then fluctuate about their respective equilibrium values $\bar{n}_s$.

For the heat current therefore the deviation of the phonon distribution (in velocity or energy) from the equilibrium value, which is produced by a temperature gradient, is decisive. For the calculation of $n_s - \bar{n}_s$ we need the Boltzmann transport equation. This equation is concerned with the particle distribution which is produced on the one hand by the interactions of the phonons with each other or with other objects (boundaries, defects, etc.), which are expressed classically by "collisions", and on the other hand by the temperature gradient. In the stationary state the total change in time of the distribution, which on the one hand originates from the collisions (index c) and on the other hand from the temperature gradient (index T), must vanish:

$$\left(\frac{\partial n_s}{\partial t}\right)_c + \left(\frac{\partial n_s}{\partial t}\right)_T = 0 \ . \tag{4.14}$$

This is already the basic form of the Boltzmann equation. It expresses the fact that in the stationary state the number of phonons $n_s(r)d^3r$ contained in the volume element d^3r at each position r must be independent on time.

The above considerations show that in formulating the transport of heat energy by lattice vibrations our picture of the phonon system has to be modified. The occupation numbers $\bar{n}_s$ and n_s are functions of temperature and therefore, in the presence of a temperature gradient, functions also of the space coordinates $r = (x, y, z)$. We thus need to think of the phonons as being *localized*, and according to the uncertainty principle this can only be done at the expense of the precision in specifying the momentum, i.e. the wave vector q. By a superposition of states, it is possible to define wave packets of wave vector range Δq, localized to within a region $\Delta x \sim 1/\Delta q$. This procedure has been outlined by *Peierls* [4.2b]. These wave packets can now be treated as localized particles moving with the group velocity v_s.

Turning back to the Boltzmann equation (4.14) we now have to find expressions for the collision term and the temperature gradient term which gives rise to a drift of the phonons. We consider first the latter term. Since the phonons are drifting in the temperature field, the same phonons are in the volume element d^3r at r at the time $t + \Delta t$ as were in the volume element at $r - \delta r = r - v_s t$ at the time t. The change in the time Δt is therefore equal to the difference in the number of phonons in these two volume elements. From this follows the differential quotient:

$$\begin{aligned}
\left(\frac{\partial n_s}{\partial t}\right)_T &= \lim_{\Delta t \to 0} \left(\frac{n_s(r - v_s\Delta t) - n_s(r)}{\Delta t}\right) \\
&\cong -v_s \nabla_r n_s(r) = -v_s \left(\frac{\partial n_s}{\partial T}\right) \nabla_r T \\
&\cong -\left(\frac{\partial \bar{n}_s}{\partial T}\right)(v_s \nabla_r T) \ .
\end{aligned} \tag{4.15}$$

Equation (4.15) gives the rate of change of the occupation number due to the motion of the phonons. This is proportional to the net flow of phonons out of a small volume element d^3r due to the variation of temperature with position. According to (4.14) this drift term is opposed by the collision term. The change in the distribution due to collisions can be attributed to the fact that, because of the change in momentum connected with a collision, phonons leave a volume element on the one hand and phonons come from neighbouring volume elements into this volume element on the other hand.

The calculation of the collision term is a complicated problem. In principle, the contribution to this term originating from phonon-phonon interactions can be calculated quantum mechanically by applying perturbation theory, considering the anharmonic terms of the potential energy as a perturbation [4.14–16]. The resulting expression for the collision term is very complicated, even if simplifying assumptions are made. A qualitative discussion is outlined in Sect. 4.5.4. At this point it should also be remembered that the mean-free paths of the phonons and hence the collision term depend not only on phonon-phonon collisions (anharmonic interactions) but also on collisions of phonons with defects and boundaries. In view of all these complications, a simple phenomenological ansatz for the collision term is adopted in the following; this ansatz is physically appealing and forms the basis for a qualitative discussion of the experimental results.

We have already mentioned above that in an anharmonic crystal any deviation of n_s from its equilibrium value $\bar{n}_s$ disappears if the temperature gradient is switched off. In order to obtain the time dependence of n_s due to collisions alone it may be assumed that n_s relaxes to its equilibrium value $\bar{n}_s$ according to an exponential law:

$$\left(\frac{\partial n_s}{\partial t}\right)_c = -\frac{n_s - \bar{n}_s}{\tau_s} \ . \tag{4.16a}$$

The left hand side denotes the time rate of change due to collisions in the general sense, including phonon-phonon, phonon-boundaries as well as phonon-defect collisions. τ_s is the *relaxation time* of phonons in the state $s = (qj)$. Since $(\partial \bar{n}_s/\partial t)_c = 0$, Eq. (4.16) can also be written in the form

$$\frac{\partial(n_s - \bar{n}_s)}{\partial t} = -\frac{n_s - \bar{n}_s}{\tau_s} \ , \quad \text{or}$$

$$(n_s - \bar{n}_s)_t = (n_s - \bar{n}_s)_{t=0}\, e^{-t/\tau_s} \ . \tag{4.16b}$$

The physical meaning of the ansatz (4.16) is the following: if a perturbed distribution n_s has arisen because of the effect of a temperature gradient, and if this temperature gradient is switched off at $t = 0$, the equilibrium distribution is re-established by collisions alone. According to (4.16b) the perturbed distribution decays with a time constant τ_s; obviously this is the time in which the deviation $n_s - \bar{n}_s$ has decreased by a factor e.

Using (4.15, 16a) the Boltzmann equation (4.14) yields

$$n_s - \bar{n}_s = -\tau_s\left(\frac{\partial \bar{n}_s}{\partial T}\right)(v_s \nabla_r T) \ . \tag{4.17}$$

The deviation $n_s - \bar{n}_s$ is obviously produced by the temperature gradient $\nabla_r T = (\partial T/\partial x_1, \partial T/\partial x_2, \partial T/\partial x_3)$; if $\nabla_r T = 0$, $n_s = \bar{n}_s$ in the equilibrium state as anticipated before. Using (4.11, 13, 17) we obtain for the heat current

$$J = -\sum_s \tau_s \hbar \omega_s \left(\frac{\partial \bar{n}_s}{\partial T}\right)(v_s \nabla_r T) v_s \ . \tag{4.18}$$

Writing $(v_s \nabla_r T) = \sum_\beta v_{s\beta}(\partial T/\partial x_\beta)$ and introducing the contribution to the specific heat of the mode s, namely

$$C_s = \hbar\omega_s \frac{\partial \bar{n}_s}{\partial T} = k_B \left(\frac{\hbar\omega_s}{k_B T}\right)^2 \frac{\exp(\hbar\omega_s/k_B T)}{[\exp(\hbar\omega_s/k_B T) - 1]^2} \ , \tag{4.19}$$

the heat current in the α-direction can be written in the form

$$\begin{aligned} J_\alpha &= -\sum_\beta \sigma_{t\alpha\beta} \frac{\partial T}{\partial x_\beta} \\ &= -\sum_\beta \left(\sum_s v_{s\alpha} v_{s\beta} \tau_s C_s\right) \frac{\partial T}{\partial x_\beta} \ . \end{aligned} \tag{4.20}$$

For the thermal conductivity $\sigma_{t\alpha\beta}$, giving the heat conduction in the α-direction produced by a temperature gradient $\partial T/\partial x_\beta$ in the β-direction, we obtain

$$\sigma_{t\alpha\beta} = \sum_s v_{s\alpha} v_{s\beta} \tau_s C_s \ . \tag{4.21}$$

If there is only a temperature gradient in the α-direction, i.e. $\partial T/\partial x_\beta = \delta_{\alpha\beta}\partial T/\partial x_\alpha$, we obtain from (4.20)

$$J_\alpha = -\sigma_{t\alpha\alpha} \frac{\partial T}{\partial x_\alpha} = -\sum_s v_{s\alpha}^2 \tau_s C_s \frac{\partial T}{\partial x_\alpha} \ ,$$

and hence

$$\sigma_{t\alpha\alpha} = \sum_s v_{s\alpha}^2 \tau_s C_s \ . \tag{4.22}$$

For an isotropic material $v_{s\alpha}^2$ may on the average be replaced by $v_s^2/3$ and (4.22) gives

$$\sigma_t = \frac{1}{3} \sum_s v_s^2 \tau_s C_s \ , \tag{4.23}$$

which is identical with (4.9). The expression (4.23) can be further simplified if we adopt the Debye model in its simplest form, in which $v_s = v$ for all modes s, that is the dispersion curves are approximated by a single acoustic branch, specified by a constant velocity v which is an appropriate average over the transverse and longitudinal branches [Ref. 4.1, Sect. 3.5.1]; in this approximation (4.23) reduces to

$$\sigma_t = \frac{1}{3} v^2 \sum_s \tau_s C_s \ . \tag{4.24}$$

If we further assume that the relaxation times are the same for all modes, i.e. $\tau_s = \tau$, which in general is a poor approximation, we obtain

$$\sigma_t = \tfrac{1}{3}v^2\tau C = \tfrac{1}{3}vlC \ , \tag{4.25}$$

where $l = v\tau$ is a mean free path and $C = \sum_s C_s$ is the total specific heat per unit volume; this relation is identical with (4.8) which gives the thermal conductivity of an ideal gas, and is often the starting point for a discussion of thermal conductivity of a solid. Its use tends to obscure the fact that phonons of different properties contribute to σ_t in different ways and to very different extents. For optical vibrations, for example, the group velocity v_s in (4.23) will usually be small since their frequency dispersions $\omega_j(\boldsymbol{q})$ are rather flat; therefore, the optical vibrations make only a small contribution to the thermal conductivity. It is rather determined by the acoustic vibrations for which the velocity changes into the sound velocity in the limiting case of long wavelengths.

4.5 Relaxation Times in Insulators

In (4.23, 24) the only quantities which are not known when the phonon dispersion $\omega_j(\boldsymbol{q})$ is known, are the relaxation times $\tau_s = \tau_j(\boldsymbol{q})$ which we shall discuss in the following. In detailed theories of the interactions of phonons with the boundaries and imperfections, the Debye model is usually introduced and in this approximation the thermal conductivity is given by (4.24), which will be the basis of most of the following discussion.

When two or more distinct scattering processes operate simultaneously, it is assumed that the effective value of the collision frequency τ_s^{-1} for the mode s is found by adding the individual collision frequencies for each process. Let $\tau_s^{-1}(\text{ph})$, $\tau_s^{-1}(\text{b})$ and $\tau_s^{-1}(\text{d})$ be the collision frequencies corresponding to phonon-phonon, phonon-boundaries and phonon-defect interactions, respectively. The effective collision frequency of mode s is than

$$\tau_s^{-1} = \tau_s^{-1}(\text{b}) + \tau_s^{-1}(\text{d}) + \tau_s^{-1}(\text{ph}) \ . \tag{4.26}$$

When one of these processes has a much higher collision frequency (shorter relaxation time) than the others it is therefore the only one which need be considered; σ_t will then be limited by this particular process.

In the following we shall give a qualitative discussion of the various scattering mechanisms and associated relaxation times which are operative in different temperature ranges.

4.5.1 Boundary Scattering

At very low temperatures only modes of low frequencies are thermally excited [Ref. 4.1, Fig. 2.10] and the mean occupation numbers $\bar{n}_s$ of these modes are very small so that the phonon-phonon collision frequencies $\tau_s^{-1}(\text{ph})$ are approximately zero. Furthermore, since the wave lengths λ_s of these modes are very large ($\lambda = 2\pi/q$, $q \ll \pi/a$), they are not scattered by localized defects, because waves with such long wavelengths do not "see" defects with dimensions of the order of a lattice parameter a. At very low temperatures the phonons are therefore only scattered by the boundaries, and the mean-free path is the same for all modes, approximately the diameter of a cylindrical specimen. The collision frequency is then given by

$$\tau^{-1} \cong \tau^{-1}(\text{b}) = \frac{v}{D} \tag{4.27}$$

and we expect (4.25) to apply with $l = D$, the diameter of the cylinder. Since in this temperature range the Debye approximation gives $C(T) = AT^3$ [Ref. 4.1, Eq. (3.108)] the theory predicts

$$\sigma_t = \tfrac{1}{3}vDAT^3 \tag{4.28}$$

at very low temperatures. The phonons produced in a heat pulse experiment will then travel "ballistically" through the crystal, with the group velocity of transverse and longitudinal modes, respectively. Figure 4.4 shows σ_t as a function of temperature for Al_2O_3 [4.17]. The three curves correspond to three specimens of different diameters prepared from the same single crystal.

Al_2O_3 has a high Debye temperature ($\theta_D \cong 1000\,\text{K}$), and $C(T) \sim T^3$ for $T \lesssim 30\,\text{K}$. For the purest crystals σ_t was indeed found to vary as T^3 below $10\,\text{K}$, and the mean free path determined from (4.25) was approximately equal to the diameter D of the specimen.

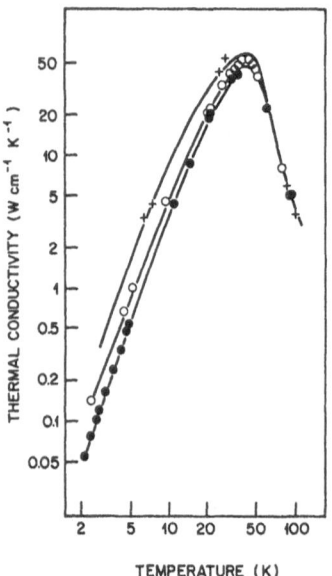

Fig. 4.4. The thermal conductivities of cylindrical specimens of Al_2O_3 having diameters 1.02 mm ($\bullet$), 1.55 mm ($\circ$) and 2.8 mm ($+$), respectively [4.17]

4.5.2 Defect and Impurity Scattering

As the temperature increases, modes having a larger value of q acquire an appreciable value of $C_j(q)$ and begin to contribute to σ_t. With increasing values of q the wavelengths of the modes decrease and become comparable to the dimensions of the defects and impurities and are therefore more and more scattered by this mechanism. The collision frequencies τ_s^{-1} are therefore approximately equal to $\tau_s^{-1}(d)$. Due to this mechanism σ_t increases more slowly with T and eventually reaches a maximum value which depends strongly on the impurity and defect concentration. For illustration, consider the case of mass defects, i.e. a crystal composed of atoms with masses m and $m + \Delta m$ arranged at random, as is the case for a crystal containing two different isotopes of the same element. In this case the theory predicts [4,18]

$$\tau_s^{-1}(d) = n_d \frac{v_a}{4\pi v^3} \left(\frac{\Delta m}{m} \right)^2 \omega_s^4 , \tag{4.29}$$

where n_d is the defect concentration, v_a the unit cell volume, v the average group velocity, and ω_s the frequency of the vibrational mode s. Note that in this case the collision frequency is proportional to ω^4, demonstrating that modes with short wavelengths are much more strongly scattered than waves with long wavelengths ($\omega = vq = 2\pi v/\lambda$ in the Debye model). Figure 4.5 shows the isotope effect on the thermal conductivity of LiF [4.10]. The

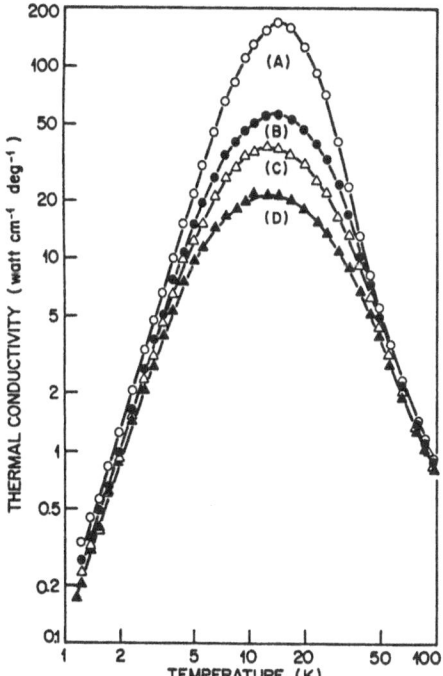

Fig. 4.5. Thermal conductivity of LiF showing the effect of isotopes. $a/o\ ^7Li$ in LiF: (A) 99.99, (B) 97.2, (C) 92.6 (natural LiF), (D) 50.8. Mean crystal widths: (A) 7.25 mm, (B) 5.33 mm, (C) 5.44 mm, (D) 5.03 mm [4.10]

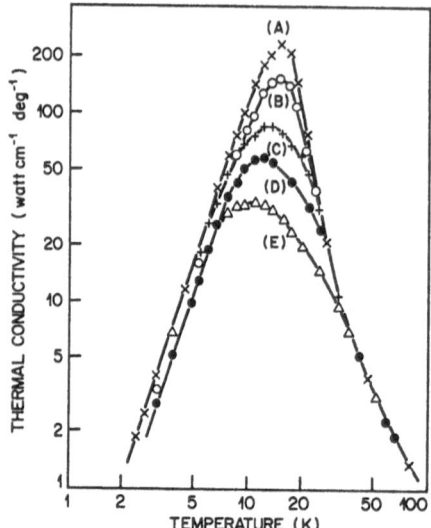

Fig. 4.6. Thermal conductivity vs temperature for NaF samples of varying purity: *(A)* best fourth-regrowth sample; *(B)* a less pure fourth-regrowth sample; *(C)* best single-growth sample; *(D)* an average single-growth sample; *(E)* best Harshaw sample [4.19]

figure shows that with increasing defect concentration the peak value of σ_t decreases considerably.

Figure 4.6 shows $\sigma_t(T)$ for very pure crystals of NaF [4.19]. This material was chosen because both Na and F occur naturally as single isotopes; the random distribution of isotopic mass need therefore not be considered in NaF. An extended multiple-growth technique was used to produce the ultrapure NaF single crystals. This technique employs repeated fractional recrystallization to lower the impurity content in the single-crystal growths. The material giving curve A had only about one part per million (1 p.p.m) of impurities, mainly Ca and K, and curve B, 6 p.p.m.

From the detailed analysis which involved estimates of relaxation times for impurity scattering and dislocation scattering and their dependence on phonon frequency, it has been concluded that if all impurities could be eliminated the peak value of σ_t would be 310 W cm^{-1} deg^{-1}, while if dislocations were absent it would rise further to 400 W cm^{-1} deg^{-1} [4.19].

Erdös and *Haley* [4.39,40] have solved the Boltzmann transport equation exactly for an insulator with random scattering centers and found that at low temperatures the thermal conductivity due to impurity scattering varies as

$$\sigma_t \sim N^{-3/4} \ ,$$

where N is the impurity concentration. This theoretical result is in agreement with experiments on the thermal resistivity $r_t = \sigma_t^{-1}$ of InSb [4.41], as shown in Fig. 4.7.

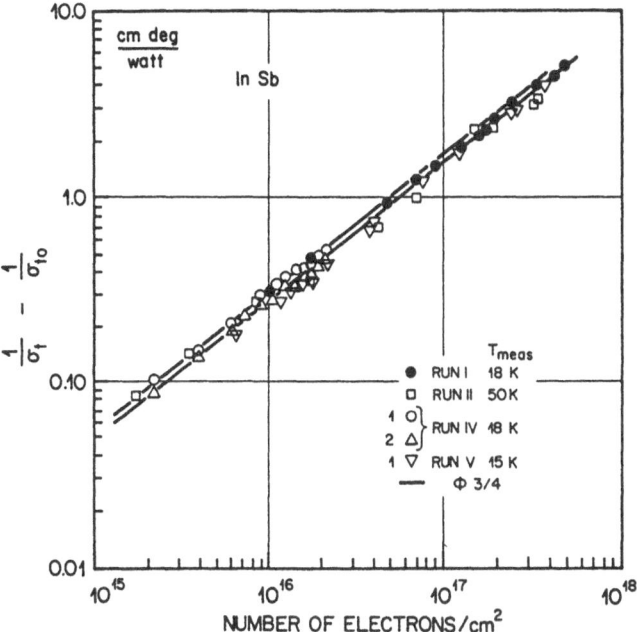

Fig. 4.7. The thermal resistivity $r_t = \sigma_t^{-1}$ of InSb containing point defects less the thermal resistivity $r_{t0} = \sigma_{t0}^{-1}$ of the sample without defects. The defects are created by electron irradiation, hence the scale of the abscissa is proportional to the defect concentration. The straight lines show the 3/4 power concentration dependencies of the data [4.41]

Another important result derived exactly by *Erdös* and *Haley* [4.39, 40] is, that at low temperatures the impurity-scattering limited conductivity is also proportional to T^3, as can be seen from Figs. 4.4–6 below about 5 K.

Since different types of imperfections lead to different forms for the frequency dependence of $\tau_s^{-1}(d)$ one can draw conclusions about the nature of lattice imperfections from the thermal resistance which they cause. Both the frequency variation and an estimate of the magnitude of the mean free path $l(\omega)$ have been obtained for a number of imperfections [4.7].

4.5.3 Phonon-Phonon Scattering

From Fig. 4.6 it is apparent that beyond about 30 K the curves for crystals of different impurity content run together again, showing that another scattering mechanism is dominant. This is the mutual scattering of lattice waves which does not occur in the harmonic approximation. It is the only intrinsic scattering mechanism because it is present even in a completely defect-free and infinitely extended crystal. Because of the complexity of the theory only a qualitative account will be given.

a) Qualitative Discussion of Microscopic Approach

From a microscopic point of view the anharmonic terms of the Hamiltonian $H' = H_3 + H_4$ [Ref. 4.1, Sect. 5.5.3] introduce interactions between three and four phonons. Scattering processes occur in which phonons are created or annihilated, the occupation numbers can be changed, and the phonon mean free paths and life times are limited to finite values. As outlined in [Ref. 4.1, Sect. 5.5.3] the problem is customarily handled by perturbation methods, regarding the anharmonic terms as responsible for the transitions between the harmonic states of the lattice. Let us consider transitions between the vibrational states of the lattice, denoting the initial state by $|i\rangle$ with energy E_i, and the final state by $|f\rangle$ with energy E_f. According to the "golden rule" of perturbation theory, the transition probability per unit time is given by

$$w = \frac{2\pi}{\hbar} |\langle f|H'|i\rangle|^2 \delta(E_f - E_i) , \tag{4.30}$$

the delta function signifying conservation of energy, $E_f = E_i$. The perturbation H' for phonon-phonon interactions is given by [Ref. 4.1, Eq. (5.149)]. For the sake of simplicity we consider only the cubic anharmonic terms:

$$H' = H_3 = \sum_{s s_1 s_2} V(s, s_1, s_2) A_s A_{s_1} A_{s_2} , \tag{4.31}$$

where

$$A_s = a^+_{-s} + a_s \tag{4.32}$$

are combinations of creation and annihilation operators and the coefficients $V(s, s_1, s_2)$ are complicated functions of the eigenfrequencies, eigenvectors, anharmonic coupling coefficients and atomic masses, and are defined in [Ref. 4.1, Appendix P]. We also have used the notation $s = (q, j)$, $-s = (-q, j)$, $s_1 = (q_1, j_1)$, $s_2 = (q_2, j_2)$, etc. Due to translational invariance of the potential energy, the coefficients $V(s, s_1, s_2)$ are proportional to a factor $\Delta(q + q_1 + q_2)$ [Ref. 4.1, Appendix P], which signifies conservation of momentum. As an example of a three-phonon process consider the process represented by the products of operators

$$a^+\binom{q}{j} a\binom{-q_1}{j_1} a\binom{-q_2}{j_2} , \tag{4.33}$$

which occurs in (4.31). This term corresponds to the annihilation of the phonons $(-q_1, j_1)$, $(-q_2, j_2)$ and the creation of the phonon (q, j). The momentum and energy conservation of this process are given by

$$q = -q_1 - q_2 + \tau \quad \text{and} \tag{4.34}$$

$$\omega_j(q) = \omega_{j_1}(-q_1) + \omega_{j_2}(-q_2) , \tag{4.35}$$

where τ is any reciprocal lattice vector including zero.

Substituting (4.31) into (4.30) it is possible to calculate the probabilities for all the individual three-phonon processes of the type (4.33). If $w(s, s_1, s_2)$ is the probability per unit time for such a process, we obtain the total change in the occupation number n_s by summing over all possible processes:

$$\left(\frac{\partial n_s}{\partial t}\right)_c = \sum_{s_1 s_2} w(s, s_1, s_2) \ , \tag{4.36}$$

which is the basis for a microscopic calculation of the collision term [4.14–16]. Inevitably the calculations are complicated and cumbersome, and we shall merely focus on the essential features, such as the implications of the conservation laws (4.34, 35) to the physics of heat conduction.

b) Normal- and Umklapp Processes

If in (4.34) $\tau = 0$ the process is an N process (normal process); if $\tau \neq 0$ we are dealing with a U process (Umklapp process) [Ref. 4.1, Sect. 5.5.3]. We shall now discuss the vital distinction between N and U processes in connection with thermal conductivitiy. Figure 4.8 illustrates N and U processes in a two-dimensional square lattice. The square in each figure represents the first Brillouin zone in the phonon q-space; this zone contains all the possible values of the phonon wave vector. Vectors q which have arrow heads at the center of the zone represent phonons absorbed in the collision process, those with arrow heads away from the center of the zone represent phonons emitted in the collision. The reciprocal vector τ as shown in Fig. 4.8b is of length $2\pi/a$, where a is the lattice constant, and is parallel to the q_x-axis. For all processes, N or U, energy is conserved, so that $\omega_1 + \omega_2 = \omega_3$. For N processes (Fig. 4.8) we have

$$q_1 + q_2 = q_3 \ , \tag{4.37}$$

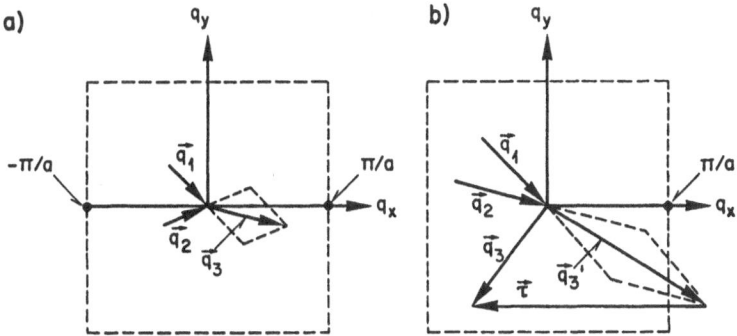

Fig. 4.8. (a) Normal (N): $q_1 + q_2 = q_3$ and (b) Umklapp (U): $q_1 + q_2 = q_3 - \tau$ phonon collision processes in a two-dimensional square lattice (*see text*); in both cases $\omega_3 = \omega_1 + \omega_2$

and the components q_{1x}, q_{2x} and q_{3x} are all positive; on the other hand, for a U process (Fig. 4.8b)

$$q_1 + q_2 = q_3 - \tau \tag{4.38}$$

and q_{1x}, q_{2x} are positive while q_{3x} is negative. This is the reason for the expression "umklapp" ("flipping" over).

Since a heat current is connected with a non-vanishing total momentum in the direction of the flow (Problem 4.9.2) we consider the effects of the N and U processes on the total momentum

$$P = \sum_{qj} n_j(q) \cdot \hbar q . \tag{4.39}$$

In thermal equilibrium, with a uniform temperature, $n_j(q) = \bar{n}_j(q) = \bar{n}_j(-q)$, and since to every vector q there corresponds a vector $-q$, P is clearly zero. In the presence of a temperature gradient, however, $P \neq 0$. Consider now the effect of a three-phonon process on P, in which the phonons $s_1 = (q_1, j_1)$ and $s_2 = (q_2, j_2)$ are absorbed and a phonon $s_3 = (q_3, j_3)$ is emitted (Fig. 4.8); after the collision process the occupation numbers are: $n'_{s_1} = n_{s_1} - 1$, $n'_{s_2} = n_{s_2} - 1$, $n'_{s_3} = n_{s_3} + 1$ and all other numbers are unchanged. From (4.39) we therefore find for the change of the momentum

$$\Delta P = P' - P = \hbar(-q_1 - q_2 + q_3) , \tag{4.40}$$

and from (4.37, 38) we obtain the result

$$\Delta P = \begin{cases} 0 & \text{for} \quad \text{N processes} \\ \hbar\tau \neq 0 & \text{for} \quad \text{U processes .} \end{cases} \tag{4.41}$$

From these considerations we see that N processes conserve the total momentum while U processes do not. Suppose we start from thermal equilibrium with a uniform temperature; in this case there is no heat flow J and $P = 0$. If a temperature gradient is switched on, P and J would remain zero if only N processes were operative. In the presence of U processes, however, ΔP changes from zero to $\hbar\tau$ if a temperature gradient is switched on and a heat flow is established. We can formulate this situation in a different way: Suppose we start from a situation in which $P \neq 0$. N processes alone could then not establish equilibrium; the energy would be redistributed among the vibrational modes but the total heat flow would remain unaltered. In fact, we should have a situation involving energy transport without any temperature gradient, implying infinite thermal conductivity. An equilibrium state without any heat current would then be impossible. We therefore reach the important conclusion that *N processes* make *no contribution* to thermal resistance. However, such processes do contribute to maintaining the "equilibrium" phonon distribution that corresponds to a steady heat flow, and

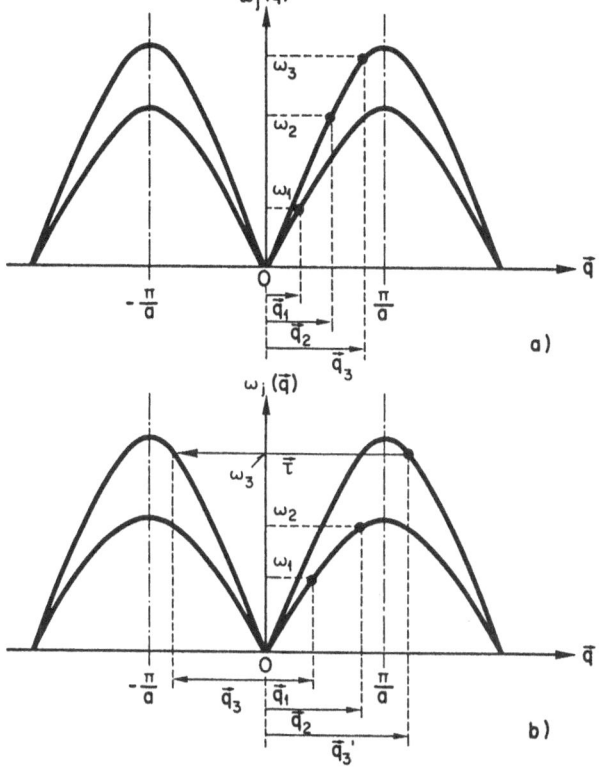

Fig. 4.9a,b. Illustration of momentum and energy conservation for (a) Normal-process (N): $q_3 = q_1 + q_2$, and (b) Umklapp-process (U): $q_3 = q_1 + q_2 + \tau$; in both cases $\omega_3 = \omega_1 + \omega_2$

their relaxation times τ_N appear in the detailed theory as well as in the theory of second-sound (Sect. 4.8).

The very different effects of N and U processes on thermal conductivity can also be illustrated qualitatively by considering the heat flow $J = \sum_s n_s \hbar \omega_s v_s$ for a one-dimensional chain as shown in Fig. 4.9. For a process in which $n'_1 = n_1 - 1$, $n'_2 = n_2 - 1$, $n'_3 = n_3 + 1$, and $n'_s = n_s$ for $s \neq 1, 2, 3$, the change in the heat current is

$$\Delta J = J' - J = \hbar \left(-\omega_1 v_1 - \omega_2 v_2 + \omega_3 v_3\right) \ . \tag{4.42}$$

Consider an N process for which q_1, q_2, q_3 are smaller than π/a and $q_1 + q_2 = q_3$, $\omega_1 + \omega_2 = \omega_3$ (Fig. 4.9a). For such a process $\omega_1 v_1 + \omega_2 v_2 \cong \omega_3 v_3$ and hence $\Delta J \cong 0$. In fact, in the Debye approximation with a mean velocity $v_1 = v_2 = v_3 = v$ we have $\Delta J = \hbar v(-\omega_1 - \omega_2 + \omega_3) = 0$. For a U process, however, with $q'_3 < \pi/a$ and $q_1 + q_2 = q_3 - \tau$, $\omega_1 + \omega_2 = \omega_3$ (Fig. 4.9b), the

95

group velocity v_3 is in the *opposite* direction to v_1 and v_2 ($v_1>0$, $v_2>0$, $v_3<0$), and hence $\Delta J<0$ or $J'<J$. In contrast to N processes, U processes are therefore very important in determining σ_t.

c) *Relaxation Times for U Processes*

It is possible to estimate the mean relaxation time for U processes, τ_u. The relaxation time is assumed to be independent of the mode and can be used in σ_t given in (4.25).

Looking at Figs. 4.8, 9b it is clear that for U processes to occur at all $q_3' = |q_1 + q_2|$ must exceed $\tau/2$. In a Debye model the radius q_D lies somewhere between $\tau/2$ and τ, and an estimate of the magnitude of the created phonon based on Figs. 4.8, 9b and energy conservation gives

$$q_3' \gtrsim \alpha q_D \quad \text{or}$$

$$\hbar\omega(q_3) = \hbar\omega_3 \gtrsim \alpha k_B \theta_D , \tag{4.43}$$

where α is a fraction in the neighborhood of $1/2$ or $2/3$. Now the collision frequency for U processes, τ_u^{-1}, is proportional to the product of the occupation numbers $\bar{n}_1 = \bar{n}(q_1)$ and $\bar{n}_2 = \bar{n}(q_2)$ of the colliding phonons. At temperatures considerably lower than θ_D one has

$$\bar{n}_1 = [\exp(\hbar\omega_1/k_B T) - 1]^{-1} \cong \exp(-\hbar\omega_1/k_B T) ,$$

and similarly for $\bar{n}_2$, and the product approximates to

$$\bar{n}_1 \bar{n}_2 \cong \exp(-\hbar\omega_3/k_B T) \cong \exp(-\alpha\theta_D/T) , \tag{4.44}$$

where we have used energy conservation and (4.43). For the collision frequency we therefore expect

$$\tau_u^{-1} \cong e^{-\alpha\theta_D/T} , \tag{4.45}$$

and on the basis of (4.25) we therefore obtain

$$\sigma_t \cong e^{\alpha\theta_D/T} . \tag{4.46a}$$

Other factors (e.g., the specific heat) would tend to modify the temperature dependence in this region, giving a function of the form

$$\sigma_t = BT^n e^{\alpha\theta_D/T} , \tag{4.46b}$$

to be fitted to the experimental results by adjusting the parameter B, n and α. For the purest crystals of NaF with $\theta_D \cong 450\,\text{K}$ [4.49] (Fig. 4.6, curve A), Eq. (4.46) was found to apply accurately between 20 and 35 K, and this is found to be the case for many materials in the region beyond the peak in the termal conductivity. Due to the factor $\exp(\alpha\theta_D/T)$ in (4.46) it also follows

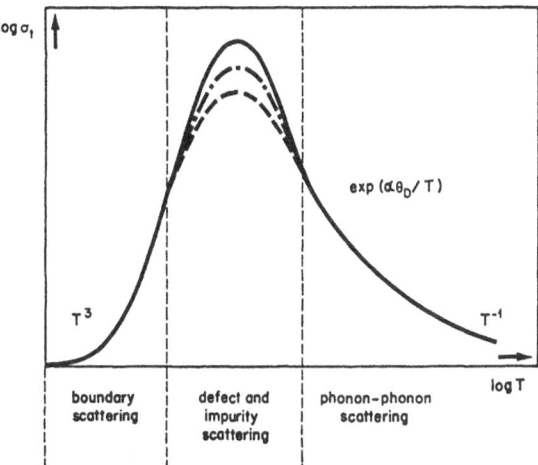

Fig. 4.10. Typical log-log plot of thermal conductivity versus temperature of insulators, showing the principal regions

that at still lower temperatures, below the peak of σ_t, the contribution to σ_t from the U processes freeze out rapidly.

At temperatures comparable or higher than θ_D, $C_s = C_j(\boldsymbol{q})$ tends to the constant value k_B for all phonons and $\bar{n}_s(T) \sim T$. Therefore, σ_t is expected to become inversely proportional to T,

$$\sigma_t \sim 1/T \ , \tag{4.47}$$

in agreement with experiment. Figure 4.10 shows a typical log-log plot of the temperature dependence of the thermal conductivity of insulators, illustrating the principle temperature regions and interaction mechanisms.

4.6 Thermal Conductivity of Glasses

Grossly disordered materials such as glasses have low thermal conductivities, one or two orders of magnitude less than those of single crystals of typical ionic materials at room temperature. Values of the latter are often in the range 0.5 to 0.1 $W\,cm^{-1}\,deg^{-1}$, very much lower than the peak values of σ_t in Figs. 4.4–6. The temperature dependence of σ_t of glasses is also qualitatively different from that of single crystals. In glasses the thermal conductivity decreases as the temperature is lowered, even at room temperature. Figure 4.11 shows the temperature dependence of vitreous silica and soda silica [4.20–22]. The mean-free path in quartz glass at room temperature is 8 Å, which is of the order of magnitude of the dimensions of a silicon dioxide tetrahedron (7 Å). A glass such as fused quartz is made up of a random but continuous network of Si-O bonds. The effective crystallite size is only of the order of a single tetrahedron of the structure. We therefore expect

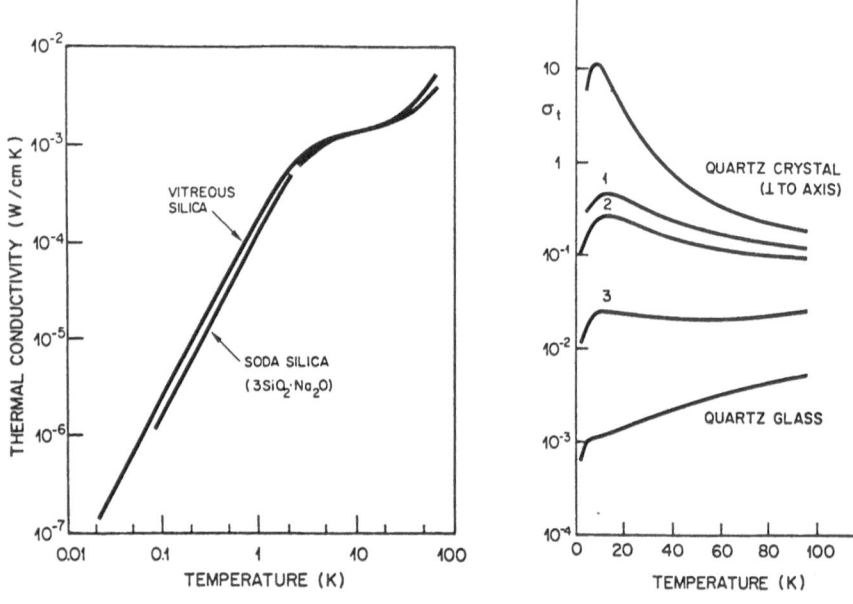

Fig. 4.11

Fig. 4.12

Fig. 4.11. Thermal conductivity of vitreous silica and soda silica as a function of temperature. σ_t varies with T^2 below 1 K and exhibits a plateau at temperatures around 10 K [4.23]

Fig. 4.12. The effect of neutron irradiation on the thermal conductivity of SiO_2. The numbers 1, 2, 3 indicate neutron irradiation for increasing periods of time [4.24]

that at sufficiently high temperatures the phonon mean free path will be constant, limited by the crystallite size, i.e. by the scale of the disorder. At lower temperatures, however, the phonon wavelengths are so long that the structure looks uniform and the mean free path is expected to increase with decreasing temperature. Applying Eq. (4.25) with $C \sim T^3$ and using the experimental fact that $\sigma_t \sim T^2$ (Fig. 4.11), a mean free path l of the dominant thermal phonons is obtained proportional to T^{-1} [4.23].

An interesting series of observations has been reported for the thermal conductivity of neutron irradiated quartz [4.24]. The neutron irradiation first creates point defects in clusters, which finally interact and lead to a general structural disorder. The thermal conductivity measured after neutron irradiation for successive periods of time is shown in Fig. 4.12. It will be seen that the conductivity decreases, that the peak due to the U processes is suppressed, and that the conductivity eventually approximates to the type found for quartz glass as shown in Fig. 4.11.

4.7 Thermal Conductivity of Metals and Alloys

The topic of the thermal conductivity in metals lies somewhat outside the scope of this book and we therefore make only a few comments. The first point to decide in discussing the thermal conductivity of metals and alloys is whether the electrons or the phonons carry the greater part of the heat current. In general there will be two contributions to the total thermal conductivity σ_t, the lattice or phonon contribution $\sigma_{t,ph}$ considered until now, and the electronic contribution $\sigma_{t,el}$. Detailed studies have shown that in normal pure metals the electrons usually carry almost all the heat current $(\sigma_{t,el} \gg \sigma_{t,ph})$, whereas in very impure metals or in disordered alloys the phonon contribution may be comparable with the electronic contribution $(\sigma_{t,ph} \cong \sigma_{t,el})$. Collisions between electrons and phonons are of great importance [4.25].

The presence of conduction electrons gives rise to an additional mechanism for phonon scattering and hence influence $\sigma_{t,ph}$. It turns out that at ordinary temperatures the collision frequency for this process is small compared with that for Umklapp scattering by phonons, so that the influence of the electrons on $\sigma_{t,ph}$ is not important. However at temperatures below the maximum in the thermal conductivity, the phonon collision frequency is substantially increased by scattering with electrons. We may therefore conclude that $\sigma_{t,ph}$ is less, and at low temperatures considerably less, than it would be in an insulator of similar dynamical properties.

The contribution of the electrons to σ_t is determined by an equation similar to (4.25), namely

$$\sigma_{t,el} = \tfrac{1}{3} v_{el}^2 \tau_{el} C_{el} . \tag{4.48}$$

The electron specific heat is small compared with the lattice specific heat, except at low temperatures. For free electrons

$$C_{el} = \frac{\pi^2}{2} \frac{k_B T}{E_F} n k_B , \tag{4.49}$$

where E_F is the Fermi energy and n the electron density. However, the velocity of the electrons, v_{el}, is two or three orders of magnitude greater than phonon velocities, since only those electrons with wavevectors near the Fermi surface contribute. For a Fermi gas

$$v_{el}^2 = v_F^2 = 2E_F/m_{el} , \tag{4.50}$$

where m_{el} is the electron mass. Substituting (4.49, 50) into (4.48) gives

$$\sigma_{t,el} = \frac{\pi^2}{3} \frac{k_B^2}{m_{el}} n T \tau_{el} . \tag{4.51}$$

The relaxation time for electron collisions, τ_{el}, is determined by interaction with the boundaries, with impurities and imperfections, with phonons and with other electrons. Detailed calculations give the result that for pure metals $\sigma_{t,el}$ is generally at least an order of magnitude greater than $\sigma_{t,ph}$ over the whole temperature range, so that the latter contribution can be neglected and

$$\sigma_t \cong \sigma_{t,el} \ . \tag{4.52}$$

Figure 4.13 shows the thermal conductivity of copper as a function of temperature [4.26]. For pure metals, therefore, both charge and heat currents are carried by the electrons alone. For a free electron metal the electronic conductivity is given by

$$\sigma_{el}(T) = \frac{n}{m_{el}}e^2\tau_{el}(T) \ . \tag{4.53}$$

Figure 4.14 shows $\sigma_{el}(T)$ of pure polycrystalline copper [4.27, 28].

At low temperatures the electron collision frequency τ_{el}^{-1} is very small since the number of phonons which can take part in the scattering is greatly reduced; hence σ_{el} is very large. With increasing temperature, however, τ_{el}^{-1} increases since the number of phonons increases, and for $T > \theta_D$ we may expect the scattering to be proportional to the square of the thermal fluctuations of the ions; hence the thermal resistivity will be proportional to T and $\sigma_{el} \sim T^{-1}$. From (4.51–53) one obtains for the ratio of thermal conductivity σ_t to electronic conductivity σ_{el}

$$\frac{\sigma_t}{\sigma_{el}} = \frac{\pi^2}{3}\left(\frac{k_B}{e}\right)^2 T = LT \ . \tag{4.54}$$

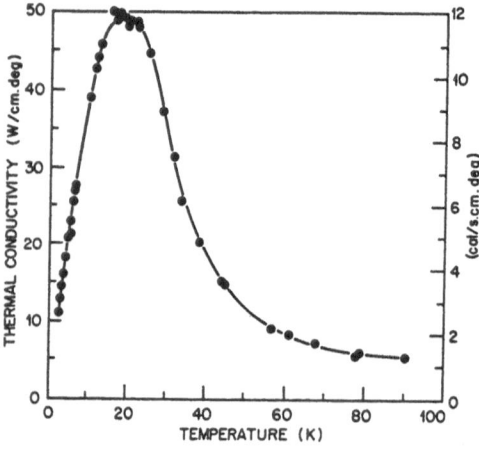

Fig. 4.13. The thermal conductivity of copper [4.26]

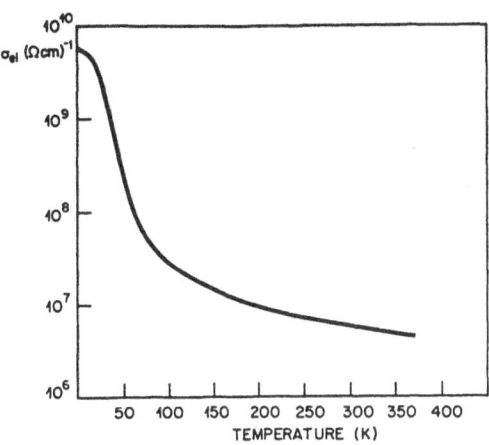

Fig. 4.14. The electronic conductivity of copper [4.27, 28]

This is the *Wiedemann-Franz law*, which states that $\sigma_t/\sigma_{el}T$ is a constant, the Lorentz number L, which depends only on the universal constants k_B and e. The value of L is 2.45×10^{-8} W Ω/deg^2. At high temperatures $(T > \theta_D)$, the experimental values of L are in good agreement with the theoretical value for many metals [4.28]. At low temperatures $(T < \theta_D)$, however, significant deviations from the Wiedemann-Franz law are observerd; the Lorentz ratio $\sigma_t/\sigma_{el}T$ is then not anymore a constant but proportional to $(T/\theta_D)^2$ [4.25].

4.8 Second Sound

We have discussed at several occasions the analogy between the phonons in a crystal and the molecules of an ordinary classical gas. One of the most striking phenomena observed in an ordinary gas is sound, a wavelike oscillatory disturbance in the local density of molecules. It is therefore natural to ask whether sound has an analogue in the phonon gas. Experiments have shown that this is indeed the case and this kind of sound propagating through the phonon gas is called *second sound*. In other words, second sound is a longitudinal phonon density wave. Second sound corresponds to an oscillation in the local phonon number density, and since the local equilibrium number of phonons is uniquely determined by the local temperature, second sound will be associated with a wavelike oscillation in temperature:

$$\delta T(\boldsymbol{r}, t) = \delta T_0\, e^{i(\boldsymbol{k} \cdot \boldsymbol{r} - \Omega t)} \ . \tag{4.55}$$

This temperature perturbation may be regarded as induced by the remainder of the phonon gas or as the result of the application of a heat pulse.

In nearly perfect crystals the only processes which do not conserve momentum are U processes, (4.41), which become very rare at low temperatures, (4.45). Under these conditions the occurrence of specific hydrodynamic phenomena as second sound and Poiseuille flow was predicted by *Sussmann* and *Thellung* [4.42] and by *Gurzhi* [4.43]. For the subject of phonon hydrodynamics in solids the reader is referred to the paper by *Thellung* [4.44] and the review article by *Beck* et al. [4.45].

In order to specify the conditions which are necessary for second sound to occur, it must be remembered that sound can propagate in an ordinary classical gas provided that:

a) Collisions between the molecules conserve number, energy and momentum.
b) The collision frequency τ^{-1} is large compared with the frequency Ω of the sound wave:

$$\tau^{-1} \gg \Omega \ . \tag{4.56}$$

Condition (b) insures that at any instant in the oscillatory cycle collisions occur rapidly enough to establish a local state of thermodynamic equilibrium. The conservation laws (condition a) are essential for the establishment of this equilibrium. The momentum conservation is of critical importance, in requiring the instantaneous local equilibrium configuration to have a nonvanishing momentum, which is the kinematical basis for the oscillation.

Now it must be recognized that the phonon gas differs in two relevant ways from an ordinary gas:

1. the phonon number is not conserved in collisions.
2. Momentum is not necessarily conserved; in fact, we have seen, (4.41), that only N processes conserve momentum but not U processes.

The violation of number conservation is not a serious problem: The loss of one of the conservation laws is related with the fact that the equilibrium phonon-distribution function (4.12) is determined entirely by the temperature, while in an ideal gas it depends on both temperature and density [4.29]. Since local equilibrium is specified by one less variable in the phonon gas, one less conservation law is required to maintain it.

The second point, however, is of more relevance, since momentum conservation is quite essential for the propagation of sound. We have seen that momentum is conserved to increasingly greater accuracy as the temperature drops, since according to (4.45) the rate of U processes, τ_u^{-1}, which destroy momentum, freeze out rapidly. The condition is that τ_u^{-1} must be small compared with the frequency Ω of the sound wave:

$$\tau_u^{-1} \ll \Omega \ . \tag{4.57}$$

In addition to this condition the analogue of (4.56) must continue to hold, where the relevant relaxation time τ_N, is that describing the momentum-conserving normal collisions, hence

$$\Omega \ll \tau_N^{-1} \ . \tag{4.58}$$

This condition guarantees that local thermodynamic equilibrium is maintained on a time scale short compared to the period of an oscillation. Combining conditions (4.57, 58) we find that the frequency must lie in the "window"

$$\tau_u^{-1} \ll \Omega \ll \tau_N^{-1} \ . \tag{4.59}$$

Second sound in the phonon gas will therefore exist at temperatures low enough for the rate of normal processes to be substantially greater than the rate of umklapp processes, and at frequencies intermediate between the two collision frequencies.

In very pure dielectric crystals the following types of heat pulse propagation are therefore expected: At very low temperatures heat pulses can propagate ballistically as transverse and longitudinal excitations with the speed of sound, also called *first-sound*. If at increasing temperature momentum-conserving phonon collisions (N processes) become more frequent, so that $\tau_N^{-1} \gg \Omega$, while the momentum-destroying collisions are still infrequent, so that $\tau_u^{-1} \ll \Omega$, the energy will propagate as a collective temperature pulse, the second sound, which travels through the crystal with a characteristic velocity. Second sound manifests itself as a wavelike oscillation in temperature as represented by (4.55). If the temperature is further increased the condition $\tau_u^{-1} \ll \Omega$ will not be valid anymore and the temperature pulses will decay by diffusion and arrive at no well defined time at the detector; this is the regime of "normal" thermal conductivity.

Second sound will usually be observed at temperature in the vicinity of the thermal conductivity maximum, between 10 and 20 K in NaF (Fig. 4.6). Since not only U processes but also impurity- or isotope scattering can be momentum destroying, second sound can only be observed in crystals of extremely high chemical and isotopic purity, such as He [4.30] or NaF [4.19, 31]. Second sound has also be observed in the semimetal bismuth [4.32]; due to the relatively low carrier concentration, the dominant mechanism for thermal conduction is via the phonons.

Figure 4.15a [4.19] shows the results of heat pulse experiments in a pure NaF sample corresponding to curve B in Fig. 4.6. The topmost trace (11 K) shows well-defined longitudinal and transverse first sound peaks, and there appears to be a slight shoulder on the decaying transverse peak. At 13 K, *three* peaks can clearly be identified, longitudinal and transverse first sound, and second sound. At 14.5 K, the second sound pulse is already superimposed on a broad diffusive ramp. Figure 4.15b [4.31] shows the heat pulses in the purest NaF sample (curve A in Fig. 4.6). The topmost trace

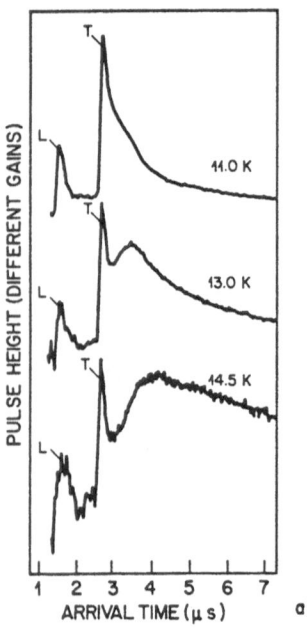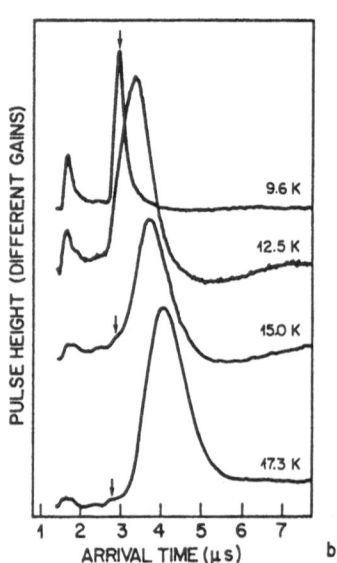

Fig. 4.15. (a) Heat pulses in a pure NaF sample ($l = 7.9\,\mathrm{mm}$) in the (100) direction for several different temperatures. L and T mark the peaks of the longitudinal and transverse ballistic pulses, respectively. Note the appearance of a third distinct pulse (second sound) [4.19]. (b) Heat pulses in the purest NaF sample ($l = 8.3\,\mathrm{mm}$) in the (100) direction for several different temperatures. The arrows mark the peak of the transverse ballistic pulse. Note the movement of the second-sound peak to later times for higher temperatures [4.31]

(9.6 K) shows well-defined longitudinal and transverse first sound pulses. At 12.5 K, the "transverse" pulse is seen to behave in an unusual fashion; it has broadened considerably and its peak has shifted to later arrival times. At 15 K a new pulse is observed which can be identified as second sound. It arrives later than the transverse pulse and has a much greater amplitude. The transverse pulse is marked by an arrow. At 17.3 K the second sound pulse has slowed enough to allow one to see the transverse ballistic pulse clearly. A comparison of Fig. 4.15b with 4.15a is instructive: At 15 K the signal in 4.15b is seen return to the base line after the second sound pulse; in contrast, at about the same temperature (14.5 K) the second sound pulse in Fig. 4.15a is already superimposed on a broad diffusive ramp. This comparison shows clearly that the purer the crystal, the greater the ratio of energy in the second sound mode to the first sound modes. It also emphasizes the very stringent conditions as to the chemical and isotopic purity of the crystal for observing second sound and explains the fact that many investigators report null results for second-sound experiments in other solids [4.30].

The discovery of second sound in superfluid ^{4}He [4.33] and in single crystals of He [4.34] has stimulated a large number of theoretical studies [4.35–37]. The frequency Ω of second sound exhibits a characteristic dispersion $\Omega(k)$ and propagates with a characteristic velocity v_{II}. If the phonon dispersion is approximated by a Debye spectrum with doubly degenerate transverse modes with velocity v_T and a longitudinal mode with velocity v_L, and if v_T is considerably smaller than v_L, the velocity of second sound is approximately given by

$$v_{II} = v_T/\sqrt{3} \; , \tag{4.60}$$

and this prediction is in agreement with experiments. This very simple result for v_{II} shows the analogy between second sound as collective mode of the phonon gas and an ordinary sound wave in a gas of particles, since in the latter case one finds the same relation between the sound velocity v_s and the mean thermal velocity v_t of a single gas particle: $v_s = v_t/\sqrt{3}$ [4.36].

4.9 Problems

4.9.1 Three-Phonon Interactions

Consider a three-dimensional crystal for which $\omega_L = v_L q$ and $\omega_T = v_T q$ where v_L and $v_T < v_L$ are the q-independent velocities of LA and TA phonons, respectively (Debye model). Using momentum and energy conservation for normal three-phonon interactions (N processes) show that the three interacting modes cannot all belong to the same branch and that the only types of processes which are allowed are the following:

TA+TA$\rightleftharpoons$LA

TA+LA$\rightleftharpoons$LA .

Note that these restrictions apply for U processes as well [4.7].

4.9.2 Relation Between Heat Flow and Phonon Momentum

Show that at very low temperatures the heat flow can approximately be written in the form

$$J = \sum_j v_j^2 P_j \; , \quad \text{where} \tag{4.61}$$

$$P_j = \sum_q n_j(q)\hbar q \tag{4.62}$$

is the total momentum for phonons of branch j and v_j is the corresponding group velocity.

Hint: Use Eq. (4.11) and the fact that at very low temperatures only modes of small frequencies are excited so that $\omega_j(q) = |v_j(\theta, \phi)|q$; for simplicity neglect the directional dependence of v_j. Note that $v_j(q)$ is parallel to q.

4.9.3 Approximate Expression for σ_t Based on Frequency Dependent Relaxation Times

Consider the expression (4.24) for σ_t which is obtained using the Debye approximation:

$$\sigma_t = \frac{1}{3}v^2 \sum_q \sum_j C_j(q)\tau_j(q) . \tag{4.63}$$

Neglecting the difference between branches and replacing the sum over the three acoustic branches $j = 1, 2, 3$ by a factor of 3 we may write

$$\sigma_t = \frac{v^2}{(2\pi)^3} \int_0^{q_D} d^3q\, C(q)\tau(q) , \tag{4.64a}$$

where we have replaced the summation over q by an integration, and where $C(q)$ is the specific heat of mode q per unit volume V. For the density of states in q-space we have used $\Gamma(q) = V/(2\pi)^3$ [Ref. 4.1, Eq. (3.44)]. Using $\omega = vq$ and $d^3q = 4\pi q^2 dq$ we can also write

$$\sigma_t = \frac{1}{2\pi^2 v} \int_0^{\omega_D} C(\omega)\tau(\omega)\omega^2\, d\omega , \tag{4.64b}$$

where ω_D is the Debye frequency and $\hbar\omega_D = k_B\theta_D$, with θ_D the Debye temperature. According to (4.26) we have

$$\tau^{-1}(\omega) = \tau_d^{-1}(\omega) + \tau_{ph}^{-1}(\omega) + \tau_b^{-1} . \tag{4.65}$$

Using (4.29) we have for defect scattering

$$\tau_d^{-1}(\omega) = A\omega^4 . \tag{4.66}$$

For phonon-phonon scattering we assume

$$\tau_{ph}^{-1}(\omega) = BT^3\omega^2 , \tag{4.67}$$

which corresponds to an approximation based on the work by *Herring* [4.38], and seems to be reasonably adequate empirically for $T < \theta_D$. For boundary scattering we have according to (4.27)

$$\tau_b^{-1} = v/D . \tag{4.68}$$

a) From (4.64–68) show that [4.16]

$$\sigma_t = \frac{k_B}{2\pi^2 v} \left(\frac{k_B T}{\hbar}\right)^3 \int\limits_0^{\theta_D/T} \frac{x^4 e^x}{(e^x - 1)^2 (Fx^4 + Gx^2 + H)} dx \;, \qquad (4.69)$$

where $F = A(k_B T/\hbar)^4$, $G = BT^3(k_B T/\hbar)^2$, $H = v/D$ and $x = \hbar\omega/k_B T$. If (4.69) is integrated analytically, it is found to be possible to fit observed $\sigma_t(T)$ curves for many materials over a substantial temperature range with reasonable choices of A, B and H.

b) If boundary scattering is ignored and if we approximate $x^2 e^x/(e^x - 1)^2$ by the leading term in its expansion for small x, show that

$$\sigma_t = \frac{k_B}{2\pi^2 v} (FG)^{-1/2} \left(\frac{k_B T}{\hbar}\right)^3 \text{arctg}[(F/G)^{1/2} \theta_D/T] \;. \qquad (4.70)$$

5. Phonons in One-Dimensional Metals

This chapter is intended to be an introduction to some physical aspects of one-dimensional (1-D) metals, in particular to their instabilities which are due to the electron-phonon interaction. We discuss the Peierls instability and the associated charge density wave (CDW), as well as the effects of the electron-phonon interaction on the phonon dispersion, i.e. the Kohn anomaly. In 1-D metals the conduction electrons can couple with internal (intramolecular) or with external (intermolecular) lattice vibrations. An example for the latter case is the quasi 1-D metal KCP which is the most thoroughly studied system. Based on the theory developed in Sect. 5.3, 4 we discuss the transport properties, diffuse X-ray and inelastic neutron scattering experiments as well as the optical properties of KCP.

5.1 Interesting Aspects of One-Dimensional Metals

Interest in the subject has grown enormously since about 1973 and has taken place primarily because highly anisotropic compounds with relatively high conductivity have become available. For reviews on the subject of quasi-one dimensional conductors the reader is referred to the literature [5.1–9].

We define a one-dimensional conductor (1-DC), as shown in Fig. 5.1: A single crystal of a 1-DC has a very high electronic conductivity parallel to a certain crystallographic axis, which we shall denote as the z-axis in the following, while the conductivity perpendicular to this axis is very

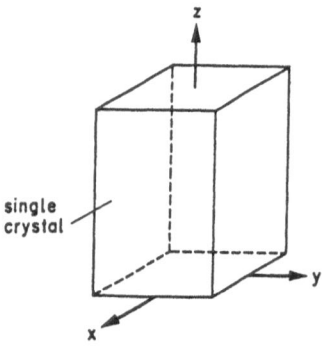

single crystal

Fig. 5.1. In a single crystal of a 1-D metal the electronic conductivity σ_{el} parallel to a crystallographic axis (the z axis) is much larger than perpendicular to this axis: $\sigma_{el}(\|z) \gg \sigma_{el}(\perp z)$

small: $\sigma_{el}(\|z) \gg \sigma_{el}(\perp z)$. Ideal 1-DC are thus crystals which show metallic behaviour in one direction and are insulators perpendicular to this direction.

There are two important conditions for a crystal to be a 1-DC: The first one is a condition for the crystal structure. The structure should be a chain structure. Within a chain the molecules or complexes should be close together so that their orbitals overlap. The parallel chains should, however, be sufficiently separated in order that the coupling between different chains is small. The second condition is a condition for the electronic band structure: There should be a partially filled conduction band. Only in this case will there be mobile charge carriers which can mover along the chain.

The study of 1-DC is interesting for several reasons:

The mathematics of one-dimensional (1-D) systems is simpler than for 2-D or 3-D systems. For 1-D systems there are models which can be solved exactly, as for instance the Ising model or the Hubbard model. For the Fröhlich model of superconductivity [5.13] an exact solution is also possible within the mean-field approximation (MFA). Experiments performed on 1-DC can therefore be compared with theoretical models which can be solved exactly.

1-DC are topologically different from 2-DC and 3-DC. It is therefore possible to study effects which are directly related to the dimensionality of the system.

Peierls [5.10, 11] has shown that the electrons in a partially filled 1-D band can always lower their kinetic energy by a symmetry reduction which splits the partially filled band into filled and empty bands. In other words, a 1-D metal is predicted to spontaneously transform into a semiconductor.

Theoreticians have worked out different models which predict superconductivity in 1-DC at non-cryogenic temperatures. Such mechanisms have been proposed by *Little* [5.12] and *Fröhlich* [5.13]. It will be discussed below why the Fröhlich superconductor does not exist in nature.

In 1-D systems thermal fluctuations play a very important role. Theoretical studies have shown that in 1-D systems, fluctuations prevent phase transitions at $T > 0$ if only short-range forces are considered.

There are three principle classes of compounds which exhibit properties characteristic for 1-DC:

Charge transfer salts of Tetracyanoquinodimethane (TCNQ) [5.1, 2, 7, 8, 14].

Conducting polymers (example: doped polyacetylene) [5.15].

Mixed valency planar transition metal compounds (example: KCP, also known as *Krogmann* salt [5.16].

The chapter is organized as follows: In Sect. 5.2 we give a qualitative discussion of some important features of 1-DC. In particular we discuss the Fermi surface, the band structure and the density of states. The Peierls instability and the associated charge density wave (CDW) as well as the effects of the electron-phonon interaction on the phonon dispersion, i.e. the Kohn anomaly [5.18] will be illustrated. Section 5.3 contains a brief theoretical description of the electronic susceptibility of a 1-DC, while in Sect. 5.4 the electron-phonon interaction is studied in its simplest form using the mean-field approximation (MFA). The results of the MFA are compared with experiments on KCP in Sect. 5.5. It turns out that the MFA is not able to account for important experimental facts due to the neglect of fluctuations which are particularly important in 1-D systems. The effects of fluctuations and three-dimensional coupling will be discussed in Sect. 5.6. While KCP is an example for the coupling of electrons with *external* or intermolecular lattice vibrations [Ref. 5.19, Sect. 4.6], *internal* or intramolecular lattice vibrations couple strongly with the conduction electrons in the charge transfer compound TEA (TCNQ)$_2$ [5.36, 5.55–57].

In the whole chapter emphasis is placed on phonons and on their interaction with the one-dimensional electronic system.

5.2 Basic Properties of One-Dimensional Conductors

In the following we review the most important properties of 1-DC which also serve to illustrate the dimensionality effects. The discussion will be qualitatively; a derviation of some of the results will be given in Sects. 5.3, 4 and in the review by *Toombs* [5.7].

Consider electrons moving in a linear monoatomic lattice with lattice constant c as shown in Fig. 5.2a. For nearly free electrons which are perturbed only weakly by the periodic potential of the ion cores, the electronic energy $E(k)$ is shown in Fig. 5.2a. The levels are occupied for $k < k_F$ and we assume a half filled band. For symmetry reasons the slope of $E(k)$ at the origin and at the boundary of the Brillouin zone is horizontal. Contrary to the two and three dimensional case this leads to corresponding singularities in the density of states $D(E)$.

Peierls has argued [5.10, 11] that in a 1-D partially filled band the kinetic energy of the electrons can always be lowered by a lattice distortion which creates a gap in the excitation energy. This is easy to visualize from Fig. 5.2a,b which shows the density of states $D(E)$ before and after the distortion has taken place. New Singularities in $D(E)$ now occur at $E = E_F \pm \Delta$, where 2Δ is the gap. The energy of the electronic system is lowered by forming the gap since the energies of the occupied levels are lowered and the energies of the unoccupied levels are raised. If the lowering of the electronic energy is larger than the increase of the elastic energy associated with the deformation of the chain, the distorted chain with lat-

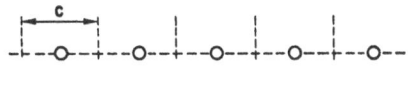

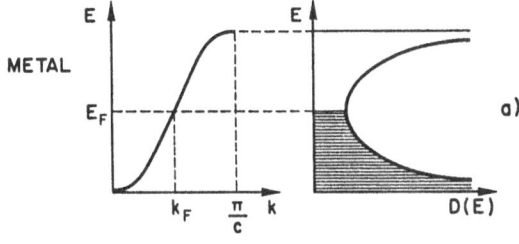

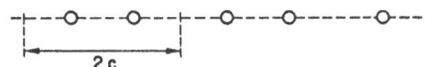

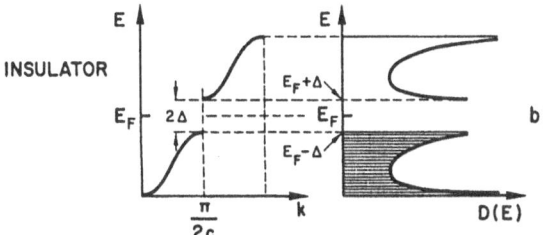

Fig. 5.2a,b. Schematic picture of the chain structure, conduction band $E(k)$ and density of states $D(E)$ of a 1-D metal (a) before and (b) after a Peierls distortion has taken place. At the transition temperature a gap 2Δ opens at the Fermi energy E_F which transforms the metal into an insulator. The figure shows the situation for a half-filled conduction band

tice constant $\lambda_P = \pi/k_F = 2c$ is more stable than the equidistant chain (Fig. 5.2b) and a transition from a metallic to an semiconducting or insulating state takes place. This transition is favoured by the singularities in $D(E)$ near E_F which strongly enhance the lowering of the electronic energy assocaited with the gap formation. It is the balance between the gain in the electronic kinetic energy and the loss in elastic energy which determines the equilibrium positions of the atoms at $T = 0$. It will be discussed in Sect. 5.4 that on the basis of mean-field (MF) theory the gap 2Δ depends on temperature (Fig. 5.7); the temperature at which the gap opens and the metal transforms into an insulator is the Peierls transition temperature T_P^{MF}.

For the half-filled band illustrated in Fig. 5.2 we have $k_F = \pi/2c$. In the general case k_F is given by

$$k_F = \frac{n}{n_{\text{max}}} k_{\text{max}} \, , \tag{5.1}$$

where $k_{max} = \pi/c$, n_{max} is the number of electrons per atom in the filled band and n is the corresponding number in the partially filled band. The wavenumber of a distortion required to produce an energy gap at k_F is $2k_F$, or more generally

$$q_P = \tau \pm 2k_F \; , \tag{5.2}$$

where τ is a suitable vector of the reciprocal lattice, i.e. $\tau = (2\pi/c)m$ with $m = 0, \pm 1, \pm 2 \ldots$. The wavelength of the Peierls distortion is then given by

$$\lambda_P = \frac{2\pi}{q_P} \; . \tag{5.3}$$

If $\lambda_P = rc$, where r is an integer, the band will be split into r subbands; those below the Fermi energy are filled and the others are empty.

In addition to the Peierls distortion, another anomaly is predicted to occur at $2k_F$ in a 1-D metal. This stems from the fact that the Fermi surface of a 1-D metal consists essentially of two points at $+k_F$ and $-k_F$. Energy conserving scattering of electrons can therefore take place exclusively from $+k_F$ to $-k_F$ and vice versa with a resultant momentum change of $2k_F$. As will be derived in Sect. 5.3 this leads to a divergence in the electronic susceptibility function $\chi(q)$ and hence in the dielectric function $\varepsilon(q)$ at $q = 2k_F$ (Fig. 5.3a). In normal 3-D metals the corresponding screening effect

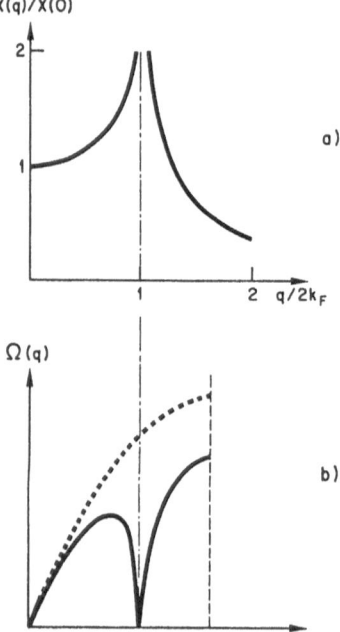

Fig. 5.3. (a) The one-dimensional susceptibility function $\chi(q)$ for zero temperature. (b) The singularity of $\chi(q)$ at $q = 2k_F$ produces a strong Kohn anomaly in the phonon dispersion of the LA mode at $q = 2k_F$

on the phonons is very weak and has been postulated by *Kohn* [5.18]; we
have given a qualitative discussion of the Kohn anomaly in 3-D metals in
[Ref. 5.19, Sect. 4.7]. In a 1-D metal, however, the effect is much stronger: if
the electrons are scattered by the phonons, the screening effect on the ions
is small if $q_{ph} \neq 2k_F$; if, however, $q_{ph} = 2k_F$, the ions are strongly screened
by the electrons, and within mean-field theory this leads to a singularity
in $\chi(q)$ at $q = 2k_F$. It is clear that this screening effect reduces the force
constant between the ions and we expect a strong Kohn anomaly in the
phonon dispersion of the LA mode at $2k_F$ (Fig. 5.3b) and possibly even
a negative eigenvalue of the $2k_F$ mode, i.e., a static Peierls distortion. The
Kohn anomaly will be studied theoretically on the basis of mean-field theory
in Sect. 5.4.

The Peierls transition of a linear chain having a partially filled conduction band induces a charge density wave (CDW) along the chain. This is
illustrated in Fig. 5.4 for the half-filled band case. In those regions where the
atomic distance is smaller than c, extra negative charge builts up, whereas
in regions of larger distances there is a lack of negative charge relative to
the equilibrium charge density of the undistorted chain.

There are two degrees of freedom associated with the $2k_F$-CDW, one describing fluctuations of its amplitude (amplitude or A_+ mode) and the other
fluctuations of its phase (phase-mode or A_- mode) (Fig. 5.5). The properties of these two collective modes will be further discussed in Sect. 5.5.4 and
the soft-mode behaviour of their frequencies as predicted on the basis of
the mean field approximation is illustrated in Fig. 5.8. *Fröhlich* [5.13] has

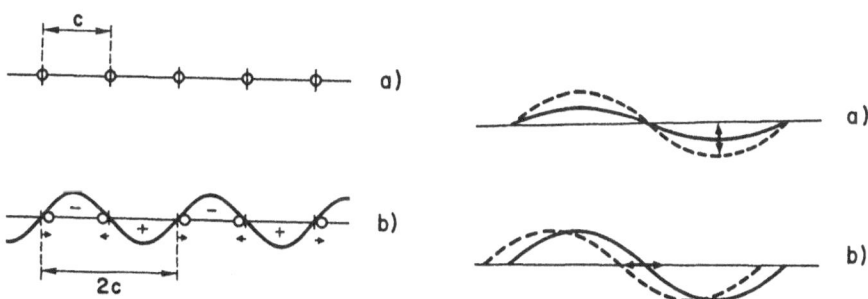

Fig. 5.4. (a) Chain of equidistant atoms of a 1-DC. (b) Schematic view of the CDW
induced by the Peierls distortion of a linear chain with a half-filled band. The Peierls
distortion leads to a dimerization of the chain with regions of negativ and positiv charges

Fig. 5.5. (a) Amplitude modulating A_+ mode of the CDW. The nodes of the CDW
remain fixed. (b) Phase modulating A_- mode of the CDW. Below the Peierls transition
temperature the $A_\pm$ modes transform into $q = 0$ modes due to the reduction of the
Brillouin zone associated with the Peierls distortion. The A_+ mode modulates the electronic polarizability and is generally Raman active, while the A_- mode induces a dipole
moment due to the displacements of charges and is therefore infrared active

argued that within a continuum model there is no restoring force acting on the CDW, i.e. the eigenfrequency of the phase-mode should be zero. As a consequence the CDW can propagate freely through the crystal under the influence of an applied field, and this would leed to superconductivity at high temperatures (Fröhlich superconductor).

Lee et al. [5.20a] have studied the Fröhlich collective phase mode in real systems. It is found that superconductivity as proposed by Fröhlich is *not possible* because in real systems there are always restoring forces acting on the phase mode which can be provided by disorder, impurities, commensurability of the CDW with the lattice or 3-D coupling between the chains [5.58]. As a result the CDW will be pinned and oscillate with a small frequency about its pinned position; this oscillation frequency is expected to be in the far-infrared (Sect. 5.5.4). Figure 5.6b shows the frequency-dependent conductivity expected for a 1-DC. The large resonance in the infrared originates from transitions across the Peierls gap, while the small peak in the far-infrared is due to the oscillation of the pinned CDW; within the Fröhlich model, the phase mode is centered at zero frequency (Fig. 5.6a).

It was suggested by *Bardeen* [5.20b] that, in addition to the oscillation about its pinned position, the CDW could also be thermally unpinned at higher temperatures and could lead to an enhanced dc conductivity. The contribution of this mechanism to the total dc conductivity is small in KCP and will be discussed in Sect. 5.5.4. In principle the CDW can not only become unpinned by thermal activation but also by an electric field [5.9]. The response of the pinned CDW to external dc and/or ac field is different from that of a simple (single particle) semiconductor, and it is anticipated that for weak pinning moderate dc electric fields can lead to a sliding CDW. Recent experiments on inorganic linear chain compounds $NbSe_3$ and TaS_3 strongly indicate that the response to external driving fields is due to the collective CDW mode and that the conductivity at high electric fields is associated with the moving charge density wave [5.9].

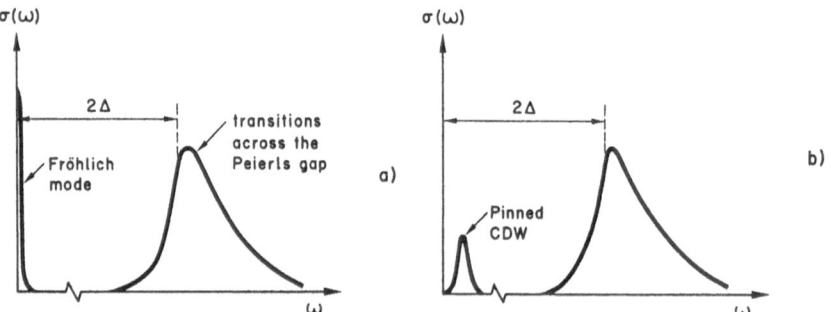

Fig. 5.6a,b. Schematic diagram of the frequency-dependent conductivity $\sigma(\omega)$ for the Peierls-Fröhlich state. (a) Without pinning of the CDW: the collective mode is centered at zero frequency. (b) Pinned CDW: the collective mode shifts to the far-infrared

On the basis of mean field theory the driving mechanism for the Peierls phase transition can be regarded as the softening of the $2k_F$ phonons at the mean field transition temperature T_P^{MF} (Fig. 5.8). The key difference to similar soft mode structural phase transitions such as the soft mode ferroelectrics (Chap. 3) is the existence of a freely propagating, current-carrying, excited state, the unpinned CDW.

It has often been pointed out that the Peierls distortion is an example of a Jahn-Teller effect. This is true to the extent that in both cases the energy is lowered by breaking the symmetry of a degenerate state. However, in the standard Jahn-Teller effect the electronic energy is linear in the distortion, while in the Peierls case it is logarithmic at $T = 0$.

5.3 The Electronic Susceptibility (By L. Pietronero)

We start with a brief description of the electronic susceptibility in a one-dimensional system for which the electronic Hamiltonian is given by [5.21]

$$H_0 = \sum_k E_k b_k^+ b_k \ . \tag{5.4}$$

Here b_k^+ and b_k are creation and annihilation operators for electronic states of energy E_k and wave vector k. It will be shown that the electronic system given by the Hamiltonian (5.4) is unstable to any arbitrarily small force of wave vector $2k_F$, where k_F is the Fermi wave vector. For this purpose we study the response of such a system to a perturbation represented by an applied scalar potential $\phi(x)$. The perturbing term of the Hamiltonian is

$$H' = \sum_j \phi(x_j) \ , \tag{5.5}$$

were x_j is the position of the j electron. It can also be written as

$$H' = \int \phi(x)\varrho(x)dx \ , \tag{5.6}$$

where $\varrho(x)$ is the electron number density operator defined by

$$\varrho(x) = \sum_j \delta(x - x_j) \ . \tag{5.7}$$

The operators which create and annihilate an electron at a given point x are

$$\psi^+(x) = \frac{1}{\sqrt{L}} \sum_{k'} b_{k'}^+ e^{-ik'x}$$

$$\psi^-(x) = \frac{1}{\sqrt{L}} \sum_k b_k e^{ikx} \ , \tag{5.8}$$

where L is the length of the system. We now obtain

$$\varrho(x) = \psi^+(x)\psi(x) = \frac{1}{L}\sum_{k,q} b^+_{k-q} b_k e^{iqx} , \qquad (5.9)$$

where $q = k - k'$. By introducing Fourier series and comparing with (5.9) we can write

$$\varrho(x) = \frac{1}{L}\sum_q e^{iqx}\varrho_q , \qquad (5.10)$$

$$\varrho_q = \sum_k b^+_{k-q} b_k , \qquad (5.11)$$

$$\phi(x) = \frac{1}{L}\sum_k \phi_k e^{ikx} . \qquad (5.12)$$

Substituting (5.10, 12) in (5.6) gives

$$H' = \sum_q \phi_{-q}\varrho_q . \qquad (5.13a)$$

In the following we consider an applied sinuisoidal potential of wave vector q for which the perturbation is given by

$$H' = \phi_q\varrho_{-q} + \phi_{-q}\varrho_q . \qquad (5.13b)$$

We compute now the change of the electronic density under the effect of the applied potential. Consider the electron-hole pair with momentum $\hbar q$ described by the operator

$$\varrho_{kq} = b^+_{k-q} b_k . \qquad (5.14)$$

The equation of motion for this operator is [5.22]

$$i\hbar\dot{\varrho}_{kq} = [\varrho_{kq}, H] . \qquad (5.15)$$

The evaluation of the commutator appearing in (5.15) proceeds as follows

$$\begin{aligned}
[\varrho_{kq}, H_0] &= b^+_{k-q} b_k \sum_{k'} E_{k'} b^+_{k'} b_{k'} - \sum_{k'} E_{k'} b^+_{k'} b_{k'} b^+_{k-q} b_k \\
&= \sum_{k'} E_{k'}(b^+_{k-q} b_k b^+_{k'} b_{k'} - b^+_{k'} b_{k'} b^+_{k-q} b_k) .
\end{aligned} \qquad (5.16)$$

From the anticommutation rules for Fermions we have [Ref. 5.21, pp. 457–461]

$$b_k b_{k'}^+ = \delta_{kk'} - b_{k'}^+ b_k$$

$$b_{k'} b_{k-q}^+ = \delta_{k',k-q} - b_{k-q}^+ b_{k'}$$

$$b_k^+ b_k^+ = b_k b_k = 0 \ . \tag{5.17}$$

Using (5.17) in (5.16) gives

$$[\varrho_{kq}, H_0] = (E_k - E_{k-q})\varrho_{kq} \ . \tag{5.18}$$

In a similar way one obtains from (5.11, 13b, 14)

$$[\varrho_{kq}, H'] = \phi_q \left[b_{k-q}^+ b_k \sum_{k'} b_{k'+q}^+ b_{k'} - \sum_{k'} b_{k'+q}^+ b_{k'} b_{k-q}^+ b_k \right]$$

$$+ \phi_{-q} \left[b_{k-q}^+ b_k \sum_{k'} b_{k'-q}^+ b_{k'} - \sum_{k'} b_{k'-q}^+ b_{k'} b_{k-q}^+ b_k \right]$$

$$= \phi_q (b_{k-q}^+ b_{k-q} - b_k^+ b_k) + \phi_{-q} (b_{k-q}^+ b_{k+q} - b_{k-2q}^+ b_k) \ . \tag{5.19}$$

The linear response of the system can be computed by taking the thermodynamic average $\langle \ldots \rangle$ of the various terms in (5.18, 19) and using the fact that for a static applied potential

$$\langle \dot{\varrho}_{kq} \rangle = 0 \ . \tag{5.20}$$

Using (5.15, 18, 19) and observing that for free electrons the thermodynamic average of the last term in (5.19) is zero, one obtains

$$(E_k - E_{k-q})\langle \varrho_{kq} \rangle + [\langle b_{k-q}^+ b_{k-q} \rangle - \langle b_k^+ b_k \rangle]\phi_q = 0 \ , \tag{5.21}$$

and application of (5.11) gives

$$\langle \varrho_q \rangle = \sum_k \frac{f_{k-q} - f_k}{E_{k-q} - E_k} \phi_q = \sum_k \frac{f_{k+q} - f_k}{E_{k+q} - E_k} \phi_q \ , \tag{5.22}$$

where we have introduced the Fermi function [Ref. 5.21, pp. 24–28]

$$f_k = \langle b_k^+ b_k \rangle = \{ 1 + \exp [\beta(E_k - E_F)] \}^{-1} \ . \tag{5.23}$$

Here, $\beta = 1/k_B T$. A useful measure of the linear density response of the electronic system to the applied potential is provided by the density response function or susceptibility

$$\chi(q) = \frac{\langle \varrho_q \rangle}{\phi_q} = \sum_k \left(\frac{f_{k+q} - f_k}{E_{k+q} - E_k} \right) \ . \tag{5.24}$$

It can be shown, that at $T = 0$ the density response function for electrons is given by

$$\chi(q) = \frac{1}{2\pi\alpha q}\ln\left|\frac{2k_F - q}{2k_F + q}\right| , \tag{5.25}$$

where $\alpha = \hbar^2/2m$, and that

$$\frac{\chi(q)}{\chi(0)} = \frac{k_F}{q}\ln\left|\frac{2k_F + q}{2k_F - q}\right| . \tag{5.26}$$

The latter function is shown in Fig. 5.3a. As can be seen $\chi(q)$ is divergent at $q = 2k_F$. As mentioned in Sect. 5.2, this singularity originates from the fact that in a one-dimensional metal with $k_B T \ll E_F$ the allowed scattering processes are only those between $-k_F$ and $+k_F$. We can therefore expand the band structure in the vicinity of $\pm k_F$,

$$E_k \cong E_F + \hbar v_F(|k| - k_F) ,$$
$$E_{k-2k_F} - E_F = -(E_k - E_F) , \tag{5.27}$$

where $v_F = \hbar k_F/m$ is the Fermi velocity, and derive a simple expression for the temperature dependence of the density response function. The result is

$$\chi(2k_F, T) \cong \frac{N(E_F)}{2}\ln(k_B T/1.14 E_F) , \tag{5.28}$$

where $N(E_F)$ is the density of states at the Fermi level. Equation (5.28) shows that $\chi(2k_F)$ diverges logarithmically with temperature. In other words, the electronic system is unstable and there is a strong tendency towards the formation of a CDW with period $2k_F$. Such a phenomenon is linked to the dimensionality of the system and disappears in two or three dimensions.

It can be seen that the instability of the free electron system to an applied potential of wave vector $2k_F$ is a generalization of the Peierls instability which produces gaps at E_F. From (5.4, 13b, 14) the Hamiltonian considered so far can be written as

$$H = \sum_k E_k b_k^+ b_k + \phi_q b_{k+q}^+ b_k + \phi_{-q} b_{k-q}^+ b_k . \tag{5.29}$$

The Hamiltonian (5.29) is a simplified form of the Hamiltonian for electrons in a rigid lattice. The potential acting on the electrons due to a lattice normally has a Fourier component for every reciprocal lattice vector. However, in the present case, the potential $\phi(x)$ is purely sinusoidal and it only mixes states with wave vectors k and $k \pm q$. The treatment of this Hamiltonian is found in any textbook of solid-state physics and the new eigenvalues of energy $\tilde{E}_k$ are given by

$$\tilde{E}_k = E_{q/2} \pm [(E_k - E_{q/2})^2 + |\phi_q|^2]^{1/2} \ . \tag{5.30}$$

The electron energy levels are split for wave vectors $k = \pm q/2$ and the potential $\phi(x)$ has created a gap of size $2|\phi_q|$ in the electronic energy spectrum. For the Peierls distortion, the potential acting on the electrons arises from the electron-phonon interaction and its periodicity is given by π/k_F or more generally by (5.2, 3). Therefore the energy gap given by (5.30) occurs at the Fermi energy as expected.

5.4 The Electron-Phonon Hamiltonian (By L. Pietronero)

The simplest Hamiltonian describing electrons in interaction with phonons is ([5.13] and [Ref. 5.21, pp. 177–182])

$$H = \sum_k E_k b_k^+ b_k + \sum_q \hbar\omega_q a_q^+ a_q + H_{\text{el-ph}} \ , \tag{5.31}$$

$$H_{\text{el-ph}} = \frac{g}{\sqrt{N}} \sum_q \varrho_{-q}(a_q + a_{-q}^+) \ , \tag{5.32}$$

where the a-operators refer to phonons [Ref. 5.19, Sect. 2.2.3], ϱ_q is the electronic density operator defined by (5.11), g is the coupling constant and N is the total number of atoms in the system. Here we refer only to the phonons of the longitudinal acoustic branch so we omit the branch index j, but in general more complex interactions can also been considered.

In order to study the effect of the coupling with electrons on the phonon frequencies we consider the equation of motion for the operator corresponding to the normal coordinates of the phonons [Ref. 5.19, Sect. 2.2.3]

$$Q_q = \sqrt{\frac{\hbar}{2M\omega_q}}(a_q + a_{-q}^+) \ , \tag{5.33}$$

where M is the atomic mass. We can write

$$i\hbar\dot{Q} = [Q_q, H] \tag{5.34}$$

and subsequently

$$i\hbar \frac{d}{dt}(i\hbar\dot{Q}_q) = -\hbar^2 \ddot{Q}_q = [[Q_q, H], H] \ . \tag{5.35}$$

It is clear that

$$[Q_q, \sum_k E_k b_k^+ b_k] = 0 \;,$$

(5.36)

so we have only to consider the phonon part of the Hamiltonian

$$H_1 = \sum_q \hbar\omega_q a_q^+ a_q \;,$$

(5.37)

and the interaction term $H_2 = H_{\text{el-ph}}$. The commutator in (5.35) can be written as

$$[[Q_q, H_1 + H_2], H_1 + H_2] = [[Q_q, H_1], H_1]$$
$$+ [[Q_q, H_1], H_2] + [[Q_q, H_2], H_1] + [[Q_2, H_2], H_2] \;.$$

(5.38)

Considering that H_2 contains the same boson operators as Q_q we have [Ref. 5.19, Sect. 2.2.3]

$$[Q_2, H_2] = 0$$

(5.39)

so that only the first two terms in (5.38) are different from zero. We have then

$$[Q_q, H_1] = \alpha \sum_{q'} \hbar\omega_{q'} [a_q a_{q'}^+ a_{q'} + a_{-q}^+ a_{q'}^+ a_{q'}$$
$$- a_{q'}^+ a_{q'} a_q - a_{q'}^+ a_{q'} a_{-q}^+] \;.$$

(5.40)

The commutation relations for bosons [Ref. 5.19, Sect. 2.2.3] imply that

$$\sum_{q'} a_{q'}^+ a_{q'} a_q = -a_q + \sum_{q'} a_q a_{q'}^+ a_{q'} \;,$$

(5.41)

$$\sum_{q'} a_{q'}^+ a_{q'} a_{-q}^+ = a_{-q}^+ + \sum_{q'} a_{q'}^+ a_{-q}^+ a_{q'} \;.$$

(5.42)

Using these relations in (5.40) we obtain

$$[Q_q, H_1] = \hbar\omega_q \left(\frac{\hbar}{2M\omega_q}\right)^{1/2} (a_q - a_{-q}^+) \;.$$

(5.43)

In a similar way we compute then

$$[[Q_q, H_1], H_1] = (\hbar\omega_q)^2 Q_q \;.$$

(5.44)

The second term in (5.38) is

$$[[Q_q, H_1], H_2] = \hbar\omega_q \left(\frac{\hbar}{2M\omega_q}\right)^{1/2} [(a_q - a_{-q}^+), H_2]$$

$$= \beta[(a_q - a_{-q}^+) \sum_{q'} \varrho_{-q'}(a_{q'} + a_{-q'}^+)$$

$$- \sum_{q'} \varrho_{-q'}(a_{q'} + a_{-q'}^+)(a_q - a_{-q}^+)]$$

$$= 2\beta\varrho_q \tag{5.45}$$

where

$$\beta = \hbar\omega_q \frac{g}{\sqrt{N}} \left(\frac{\hbar}{2M\omega_q}\right)^{1/2} . \tag{5.46}$$

This leads to the final result

$$\ddot{Q}_q = -\omega_q^2 Q_q - g\left(\frac{2\omega_q}{\hbar N M}\right)^{1/2} \varrho_q . \tag{5.47}$$

Introducing the mean-field approximation

$$\varrho_q \cong \langle\varrho_q\rangle \tag{5.48}$$

and making the identification

$$g\left(\frac{2M\omega_q}{N\hbar}\right)^{1/2} Q_q \equiv \phi_q \tag{5.49}$$

we have

$$\langle\varrho_q\rangle = \chi(q)\phi_q \tag{5.50}$$

and therefore

$$\ddot{Q}_q = -\omega_q^2 Q_q - \frac{2g^2\omega_q}{N\hbar}\chi(q) . \tag{5.51}$$

The new dispersion relation of the phonons interacting with the electrons in the mean field approximation is then given by

$$\Omega^2(q) = \omega_q^2 + \frac{2g^2\omega_q}{N\hbar}\chi(q) . \tag{5.52}$$

From (5.25) it is seen that for $T = 0$ the electronic suceptibility $\chi(q)$ is negative and diverges for $q \to 2k_F$, leading to a negative eigenvalue $\Omega^2(2k_F)$ and hence to an instability of the system. For finite temperatures $\chi(q,T)$ also becomes large and negative for $q \to 2k_F$ when the temperature is lowered. For $q = 2k_F$ we can insert (5.28) into (5.52) and obtain the temperature dependence of the phonon frequency at $2k_F$:

$$\Omega^2(2k_F, T) = \omega_{2k_F}^2 - \left\{ [g^2 N(E_F)\omega_{2k_F}]/N\hbar \right\}$$
$$\times \ln(1.14\, E_F/k_B T) \ . \tag{5.53}$$

This expression shows that by lowering the temperature the frequency at $2k_F$ decreases until it vanishes at the mean field Peierls transition temperature $T = T_P^{MF}$, given by

$$k_B T_P^{MF} = 2.28\, E_F e^{-1/\lambda} \ , \quad \text{where} \tag{5.54}$$

$$\lambda = \frac{g^2 N(E_F)}{N\hbar\omega_{2k_F}} \ . \tag{5.55}$$

The $2k_F$-phonon frequency can be written in terms of T_P^{MF} as

$$\Omega^2(2k_F, T) = \lambda\omega_{2k_F}^2 \ln(T/T_P^{MF}) \ . \tag{5.56}$$

The temperature T_P^{MF} corresponds to the Peierls instability in the mean field approximation. The fact that the phonon frequency vanishes corresponds to a static distortion of the lattice with wave vector $2k_F$. A schematic picture of the modification of the phonon spectrum below T_P^{MF} is indicated in Fig. 5.3b. We have already obtained the result that according to (5.30) the amplitude $2|\phi_{2k_F}|$ corresponds to the Peierls gap 2Δ (Fig. 5.2). Using (5.49) it is therefore possible to express the Peierls gap directly in terms of the amplitude of the statice distortion, $\overline{Q}_{2k_F}$:

$$\Delta = g(2M\omega_{2k_F}/N\hbar)^{1/2}\overline{Q}_{2k_F} \ . \tag{5.57}$$

The value of Δ is a function of temperature that can be obtained by minimizing the total static energy [5.23]. It is interesting to note that $\Delta(T)$ follows a BCS type of temperature dependence and vanishes at T_P^{MF} given by (5.54). The temperature dependence of $\Delta(T)/\Delta(0)$ is shown in Fig. 5.7. Fröhlich found that the gap at $T = 0$ is given by [5.13, 23]

$$2\Delta(0) = 8E_F e^{-1/\lambda} \ , \tag{5.58}$$

and from (5.54) it follows that

$$\frac{2\Delta(0)}{k_B T_P^{MF}} = 3.51 \ . \tag{5.59}$$

The temperature dependence of the frequencies of the amplitude A_+ mode and the phase A_- mode of the CDW as predicted by the mean-field approximation is shown schematically in Fig. 5.8. Above T_P^{MF}, the A_+ and the A_-

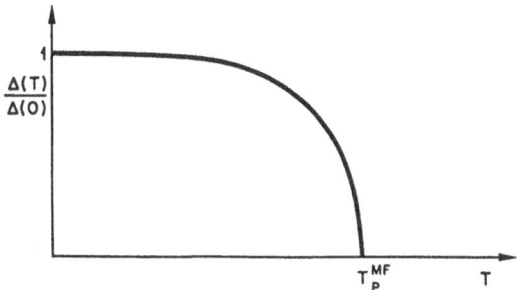

Fig. 5.7. The dependence of the
energy gap 2Δ on temperature
in the mean field approximation

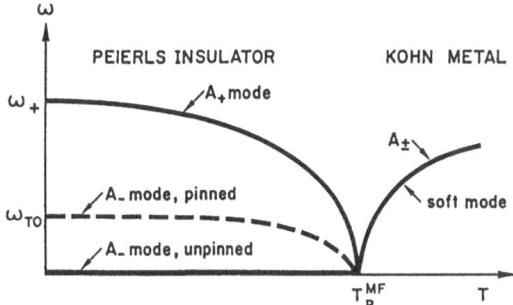

Fig. 5.8. Temperature dependence of the amplitude modulating A_+ mode and the phase
modulating A_- mode as predicted
by the mean field theory (see text)

modes are degenerate, but below T_P^{MF} the lowering of the symmetry due to
the Peierls distortion lifts this degeneracy. The frequency of the A_+ mode is
expected to increase with decreasing temperature. The frequency of the A_-
mode would be zero for vanishing pinning and is the current carrying mode
considered by Fröhlich (Fig. 5.6a). Pinning of the CDW leads to a frequency
$\omega_{TO}\neq0$ and gives rise to the far-infrared response shown in Fig. 5.6b. The
dispersion relation of the Fröhlich collective mode with frequency ω_F for the
distorted system, i.e. at $T\ll T_P^{MF}$, in the vicinity of $2k_F$ can be computed
using Green's function methods [5.24, 25]. The result is

$$\omega_F^2(\tilde{q}) = \tfrac{1}{4}\lambda\omega_{2k_F}^2\left(\tilde{q}v_F/\Delta\right)^2 , \quad \text{where} \tag{5.60}$$

$$\tilde{q} = q - 2k_F . \tag{5.61}$$

For $\tilde{q} = 0$ the result $\omega_F(\tilde{q} = 0) = 0$ gives the Fröhlich CDW superconductivity (Figs. 5.6a, 8). A phenomenological way to take pinning into account
is to write the dispersion relation of the pinned phase mode of the CDW as
[5.25]

$$\omega_-^2 = \omega_{TO}^2(\tilde{q}) = \omega_{pi}^2 + \omega_F^2(\tilde{q}) , \tag{5.62}$$

where ω_{pi} is the pinning frequency. The frequency $\omega_{TO}(\tilde{q} = 0) = \omega_{pi}$ can
be observed in the far infrared (Fig. 5.6b, 17, 18).

The effective mass m^*_{CDW} of the CDW is given by [5.20a]

$$m^*_{CDW}/m_e = 1 + \frac{4\Delta^2}{\lambda \omega^2_{2k_F}} \ , \tag{5.63}$$

where m_e is the "bar" electron mass. Typical values are $m^*_{CDW}/m_e \cong 10^3$ (Sect. 5.5.4); thus m^*_{CDW} is much less than the atomic mass M. This indicates that the steep dispersion around $q = 2k_F$ is due to the relatively small effective mass of the CDW condensate, rather than to a strong elastic energy.

Associated with the transversely polarized phase mode is a longitudinal mode whose frequency ω_{LO} is given by [5.20a]

$$\omega^2_{LO} = \tfrac{3}{2} \lambda \omega^2_{2k_F} \ . \tag{5.64}$$

The frequency of the amplitude modulating A_+ mode is found to be given by [5.26]

$$\omega^2_+ = \lambda \omega^2_{2k_F} = \tfrac{2}{3} \omega^2_{LO} \ . \tag{5.65}$$

5.5 Discussion of Selected Experiments for KCP

5.5.1 Chemistry and Structure

KCP is an abbreviation for the *Krogmann* salt [5.16]

$$K_2[Pt(CN)_4]Br_{0.3} \cdot 3H_2O \ .$$

The starting material for the preparation of KCP is $K_2[Pt(CN)_4] \cdot xH_2O$. Figure 5.9a shows the planar $Pt(CN)_4$ complexes and the d^2_z orbitals of the Pt ions with 2 electrons. The $Pt(CN)_4$ complexes form linear chains in this compound as shown in Fig. 5.9b. Successive complexes are twisted by an angle of 45°; this reduces considerably the Coulomb repulsion between the CN ligands of neighbouring complexes, resulting in relative short Pt-Pt distances of 3.2 Å in $K_2[Pt(CN)_4] \cdot xH_2O$. Therefore, the d^2_z orbitals of neighbouring Pt ions overlap and form a *filled* one-dimensional d^2_z-band, and hence the compound $K_2[Pt(CN)_4] \cdot xH_2O$ is a perfect insulator. It is, however, possible to extract some of the electrons from the filled d^2_z-band by partial oxidation with Br, so that a *partially filled conduction band* results. This is achieved by reacting 5 parts of the unoxidized compound $K_2[Pt^{2+}(CN)_4]$ with one part of the fully oxidized compound $K_2[Pt^{4+}(CN)_4]Br_2$ in aqueous solution. The product of this reaction is KCP. In KCP the Pt-Pt distance

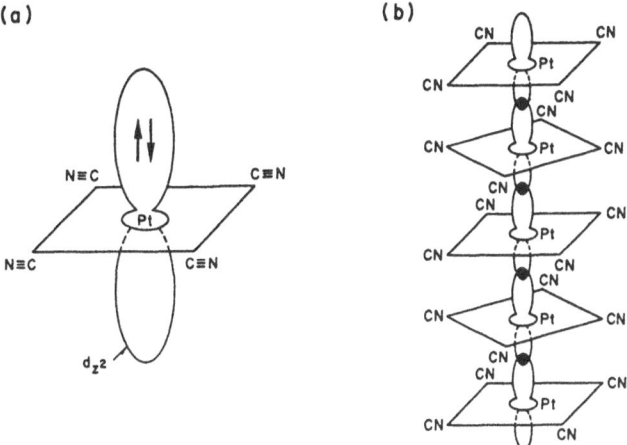

Fig. 5.9. (a) Planar Pt(CN)$_4$ complex with d_z^2 orbital which can accomodate 2 electrons per Pt atom. (b) Chain of Pt atoms in KCP showing the overlap of d_z^2 orbitals between stacked complex ions. The Pt(CN)$_4$ complexes are staggered to reduce the Coulomb repulsion between ligands [5.17]

is reduced from 3.2 to 2.87 Å which is only slightly larger than the Pt-Pt distance in the Pt-metal. The greatly increased overlap of the d_z^2 orbitals leads to an increase in the dc conductivity of about 8 orders of magnitude as compared with the starting compound. In KCP the Pt ions have a mixed valency of 2.30. It should be emphasized, however, that there is no statistical distribution of Pt^{2+} and Pt^{4+} ions but that the electrons are delocalized, as concluded from Mössbauer experiments [5.27]. The Br ions and the H$_2$O molecules are located between the parallel Pt chains and the Br content is non-stoichiometric but fixed. Since each Pt ion contributes 2 electrons to the d_z^2 band and each of the 0.30 Br per Pt extracts one electron from the d_z^2 band, there are 1.70 electrons left, resulting in an approximately 5/6 filled d_z^2 band. The electrons can move easily along the parallel Pt chains but they can hardly change from one chain to the other which explains the extreme anisotropic electrical and optical properties of this unique compound.

5.5.2 Transport Properties, Diffuse X-Ray Scattering and Band Structure

The temperature dependence of the dc conductivity $\sigma_\parallel$ and $\sigma_\perp$ parallel and perpendicular to the Pt chains observed by *Zeller* and *Beck* [5.28] is illustrated in Fig. 5.10. At room temperature, $\sigma_\parallel$ is about 5 orders of magnitude higher than $\sigma_\perp$ and has a value of several hundred $(\Omega\,\text{cm})^{-1}$. With decreasing temperature, the conductivity decreases. At 35 K, KCP seems to be an insulator, and the anisotropy $\sigma_\parallel/\sigma_\perp$ has decreased to about 10^3.

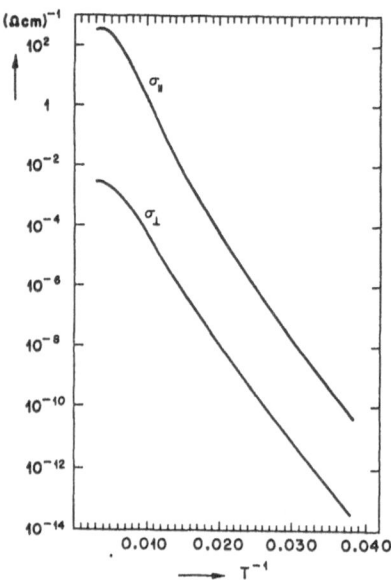

Fig. 5.10. Temperature dependence of the longitudinal and transversal dc conductivity of KCP [5.28]

Experimentally the Peierls distortion has been demonstrated for the first time for KCP by *Comes* et al. [5.29] by means of diffuse X-ray scattering. Investigating a room temperature sample with the "monochromatic Laue technique" [Ref. 5.30, Fig. 5.2], these authors found superlattice structure planes in reciprocal space. Intensity planes in reciprocal space correspond to a one-dimensional periodic structure in real space. The observed satellites demonstrate the existance of a superstructure of the Pt ions in the chain direction with a period of 6.7 Pt-Pt distances. This is exactly the period one would expect for the Peierls distortion on the basis of the chemical band filling according to (5.1–3): with $n = 1.70$, $n_{max} = 2$ we have $k_F = 0.85\pi/c$, $q_P = 2\pi/c - 1.70\pi/c = 0.3\pi/c$ and $\lambda_P = 2\pi/q_P = 6.66c$, corresponding to approximately a 5/6 filled d_z^2 band.

It might be surprising that a Peierls distortion is already observed at room temperature in the quasi-metallic state. It should be mentioned, however, that it is difficult to decide from the diffuse X-ray results alone whether the superstructure is of static or dynamic nature. It could be caused also by a lattice vibration with a wavelength of 6.7 Pt-Pt distances. If the frequency of this particular phonon is small compared to the phonon frequencies for neighboring wavevectors, the diffuse X-ray intensity should be especially large because the cross section for scattering of X-rays by phonons of frequency ω is proportional to ω^{-2} [Ref. 5.30, Eq. (5.4.2)]. In other words, it is not possible to decide whether there is a static or fluctuating Peierls distortion at room temperature. It will become clear later (Sects. 5.5.3, 4) that at room temperature the Peierls distortion is in fact fluctuating, implying a fluctuating Peierls energy gap. The X-ray data do show, however, that at room temperature the superstructure between different chains is completely uncorrelated. At low temperatures the diffuse sheet satellites concentrate more and more on particular points. This indicates that the superstructure on adjacent chains is now correlated and 180° out of phase (Fig. 5.19).

It is well known that in a conventional solid, d bands are usually narrow with large effective masses and hence far away from free-electron behaviour. At first sight it would therefore seem that due to the relatively small overlap in the d_z^2 orbitals a tight-binding model for the band structure should be a very good approximation. This is not born out by experiments performed on KCP. Experiments show that except for the Peierls gap in the vicinity of E_F and a weak pseudopotential which creates an energy gap at the boundaries of the Brillouin zone, the conduction electrons in KCP behave as nearly free electrons in the z-direction. This is concluded from the optically deduced effective mass $m^* \cong m_e$ (Sect. 5.5.4), from the oscillator strength sum rule and from energy and oscillator strength of the first interband transition. The nearly free electron band structure of KCP is shown in Fig. 5.11 [5.31]. The fundamental reason for this at first sight surprising behaviour is the following: In terms of wave functions "free electrons" means constant electron density or in our specific case an electron density independent of z, the conducting axis (Remember that for the electrons $\psi(z) \sim \exp(\mathrm{i}kz)$ and hence $|\psi|^2 = $ const.). In three dimensions a constant electron density can only be approximated by sufficient overlap of spherically symmetric s functions but never with directed d bonds. In one dimension, however, we

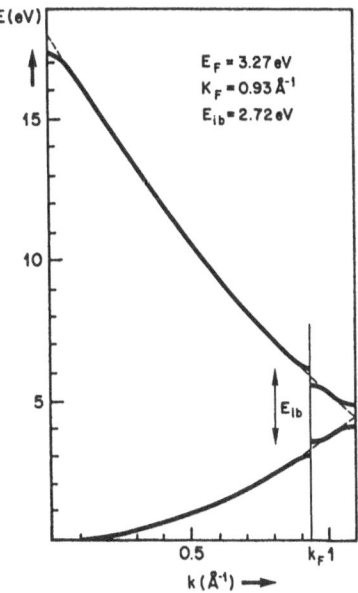

$E_F = 3.27\,\text{eV}$
$K_F = 0.93\,\text{Å}^{-1}$
$E_{ib} = 2.72\,\text{eV}$

Fig. 5.11. Nearly free electron band structure of KCP. Except for the pseudopotentials creating the Peierls gap and the gap at the zone boundaries the model has no free parameters [5.31]

127

just ask for constant electron density along the conducting axis. This can be approximated nearly as well by s functions and by s hybridized d_z^2 orbitals. In KCP and related salts the *three*-dimensional band structure is highly non-free-electron-like with formally a huge effective mass in the transverse direction. The above dimensionality argument just states that it is easy to achieve nearly free electron like behaviour in one direction and that in this respect there is little difference between an s and a hybridized d_z^2 orbital. These conclusions have been confirmed by band structure calculations [5.32, 33].

5.5.3 Inelastic Neutron Scattering Studies

The existence of a giant Kohn anomaly in KCP at room temperature has been demonstrated by *Renker* et al. [5.34]. They measured the longitudinal acoustic (LA) phonon dispersion in the direction $[0, 0, \xi]$ of the platinum chains by means of inelastic neutron scattering. Figure 5.12 shows the result of this experiment. The pronounced anomaly in the LA phonon branch of KCP is at $q_P = 0.6\pi/2c$, i.e. at the same value where the one-dimensional superlattice structure has been observed in diffuse X-ray scattering experiments [5.29]. The qualitative features of the observed Kohn anomaly are the same as illustrated in Fig. 5.3 and can be understood on the basis of the Fröhlich Hamiltonian (5.31, 32). The data of Fig. 5.12 have been measured on KCP crystals containing ordinary crystal water. The incoherent neutron scattering of hydrogen makes phonon measurements and even more so quasi-elastic studies rather difficult. Figure 5.13 shows a selection of phonons measured on deuterated KCP at room temperature [5.35]. Besides the LA phonons in the chain direction $(0, 0, \xi)$, LA phonons at the zone boundary $(\pi/a, \pi/a, q)$ of the Brillouin zone are presented. The phonon anomaly appears also at $(\pi/a, \pi/a, 2k_F)$ and was further measured at several other

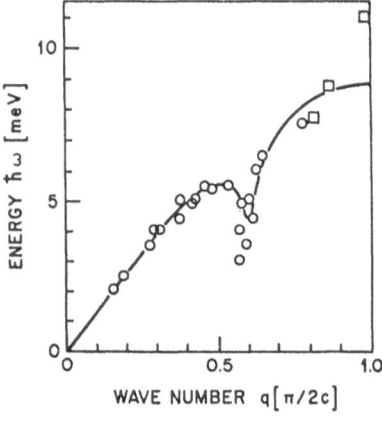

Fig. 5.12. Longitudinal acoustic phonon branch in the chain direction of $K_2[Pt(CN)_4]Br_{0.3} \cdot 3H_2O$. (—) is a model fit discussed in [5.34]

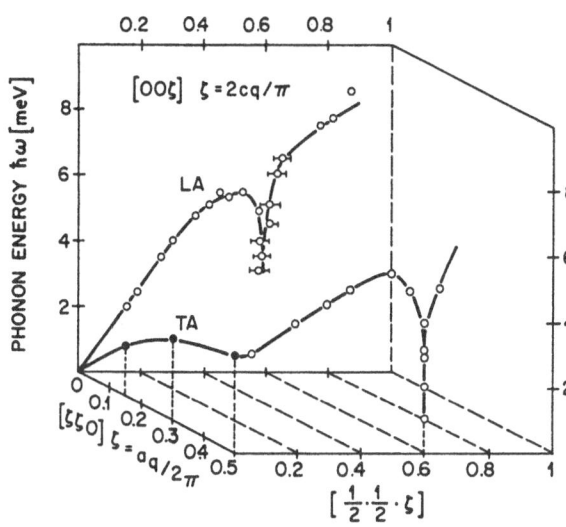

Fig. 5.13. Acoustic phonon modes measured in $K_2[Pt(CN)_4]Br_{0.3} \cdot 3D_2O$ at room temperature [5.35]

Acoustic modes in $K_2Pt(CN)_4Br_{0.3}3D_2O$ at 295 K

q-points in the $2k_F$-plane of the reciprocal lattice, which proved in a rather direct way that this effect is due to a one-dimensional property of the crystal.

The fact that at room temperature a Peierls distortion (diffuse X-ray scattering) and a Kohn anomaly (inelastic neutron scattering) have been observed, suggests that the mean-field transition temperature T_P^{MF} as given by (5.54) is considerably higher than 300 K in KCP. In fact we shall find that $T_P^{MF} \cong 600\text{–}700$ K (Sect. 5.5.4). Since KCP becomes unstable above about 350 K the transition temperature lies outside the stability region. At all experimentally accessible temperatures ($T \leq 300$ K), one is sufficiently below T_P^{MF} for $2\Delta(T) \cong 2\Delta(0)$ to hold. At room temperature the gap is fluctuating and only at low temperatures ($T < 100$ K) will the gap become static (Sect. 5.6), but the mean gap will be roughly temperature independent below 300 K. The softening of the degenerate $A_\pm$ modes above T_P^{MF} as described by (5.56) and illustrated in Fig. 5.8 is therefore outside the range of observation, and below 300 K we expect that the gap as well as the frequencies of the $A_\pm$ mode and the pinned A_- mode are roughly temperature independent.

Further inelastic neutron scattering experiments by *Comes* et al. [5.35] concluded that only at higher temperatures did the $2k_F$ phonons soften appreciably, whereas at low temperatures the Kohn anomaly had almost disappeared. These results are in contradiction with the picture emerging from more recent inelastic neutron scattering experiments performed by *Lynn* et al. [5.37] and by *Carneiro* et al. [5.38], who observed a very pronounced Kohn anomaly down to at least 80 K. Although neutron scattering

is in principle excellently suited to study lattice dynamics, the rapid variation of the excitations around the $2k_F$ anomaly in KCP makes it impossible
to measure phonon dispersion in the ordinary sense by this technique. This
may also be the reason for the discrepancy in the results and conclusions
mentioned above. The paper of *Carneiro* et al. also contains a discussion
of their results in relation to those obtained from far infrared and Raman
experiments. It is found that the inelastic neutron scattering results agree
satisfactory with the optical studies. As will be discussed in the next section the excitations observed by far infrared and Raman techniques can
successfully be interpreted as excitations arising from the pinned CDW.

5.5.4 Optical Properties of KCP

Figure 5.14 shows reflectivity measurements of KCP single crystals by *Brüesch*
and *Zeller* [5.39]. The reflectivity has been observed at room temperature
and at near-normal incidence in the wavelength region from the far-infrared
(FIR) to the ultraviolet (UV) for light polarized parallel $(E\|z)$ and perpen-
dicular $(E\perp z)$ to the conducting direction z. The reflectivity R is extremely
anisotropic. For $E\perp z$ the material behaves as a transparent dielectric: R is
small throughout the spectrum and virtually wavelength independent in the
UV and infrared (IR) region. In the FIR some phonon structure appears.
For $E\|z$, R is large throughout the FIR and IR, shows a pronounced plasma
edge near $\omega_p = 16000\,\mathrm{cm}^{-1}$ ($\sim 2\,\mathrm{eV}$) and slowly increases towards the UV.
The plasma frequency is given by

$$\omega_p^2 = \frac{4\pi e^2 n}{\varepsilon_\infty m^*} , \tag{5.66}$$

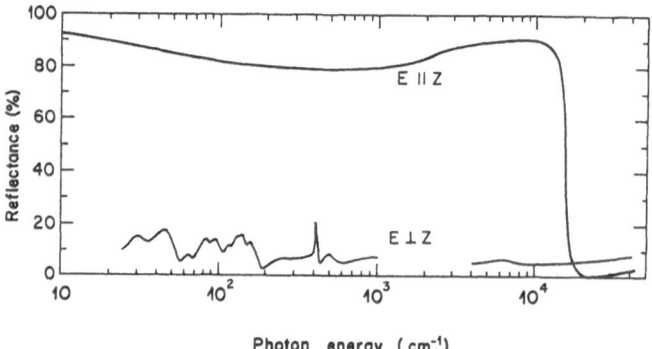

Fig. 5.14. Near-normal incidence reflection spectra of KCP measured at room temper-
ature with the light polarized parallel $(E\|z)$ and perpendicular $(E\perp z)$ to the highly
conducting z axis [5.39]

where n is the electron concentration, ε_∞ the optical dielectric constant and m^* the effective mass. With $n = 1.70$, $\varepsilon_\infty = 2.20$, $\omega_p = 2\,\mathrm{eV}$ one obtains $m^* = m_e$. This means that for $\omega > \omega_p$ the electrons behave as free electrons. For $\omega < \omega_p$ there is already some localization of the electrons at 300 K as can be seen from Fig. 5.14: For completely free electrons, R would increase continuously to 100% with decreasing frequency. In contrast the observed reflectance passes through a broad maximum of 90% in the NIR, than decreases to 78%, the decrease being most pronounced in the IR and only below $200\,\mathrm{cm}^{-1}$ increases again. This clearly shows a deviation from the Drude behaviour of free electrons and gives rise to a resonance in the frequency dependent conductivity $\sigma(\omega)$ centered around $\omega_g = 1600\,\mathrm{cm}^{-1}$ (Fig. 5.16).

Figure 5.15 shows the $E\|z$ reflectivity at 300 K and at 40 K [5.40]. At 40 K a strong structure is observed in the FIR which will be discussed below. The plasma edge is nearly temperature independent and the decrease of R in the IR towards lower frequencies is even more pronounced at 40 K than at 300 K.

The temperature dependence of the optical conductivity σ in the infrared as obtained from a Kramers-Kronig analysis of the reflectivity data is shown in Fig. 5.16. The question arises as to the origin of the infrared resonance in $\sigma(\omega)$ shown in Fig. 5.16. The strong Kohn anomaly observed at 300 K (Figs. 5.12, 13) suggests that it is the electron-phonon interaction which gives rise to the room-temperature structure in $\sigma(\omega)$ in Fig. 5.16. In fact, *Rice* and *Straessler* [5.41] have shown that the scattering of the electrons by the $2k_F$ phonons leads to a modified Drude behaviour with a frequency dependent scattering time, resulting in a mobility gap. This explana-

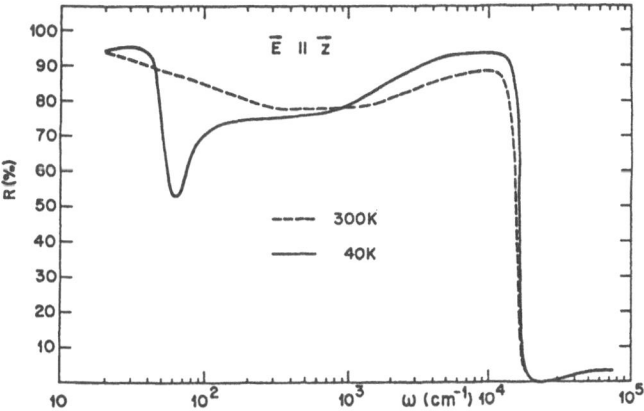

Fig. 5.15. Far-infrared to UV spectrum of KCP for light polarized parallel to the conducting axis at 300 K (- - -) and at 40 K (—). The most drastic change in the two spectra is the strong structure in the far-infrared at 40 K. This structure is assigned to the oscillation of the pinned CDW induced by the Peierls distortion [5.40]

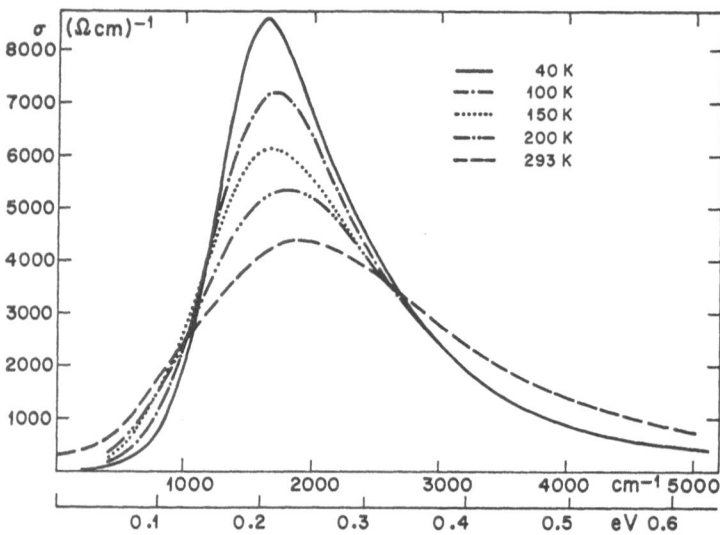

Fig. 5.16. Temperature dependence of the optical conductivity σ in the infrared. The resonance at high temperatures originates from a strong mobility pseudogap which transforms into the real Peierls gap as $T \to 0$ without significantly changing the gap energy. The resonance at 40 K is due to transitions across the Peierls gap [5.40]

tion implies that contrary to the mean field result (5.7) the peak should not shift with decreasing temperature. As already mentioned in Sect. 5.5.2, the Kohn anomaly observed at room temperature implies a fluctuating Peierls gap, and only at low temperatures a static Peierls gap will be formed. Due to the fluctuations the optical gap is expected to be roughly temperature independent. At room temperature one is therefore observing transitions across the fluctuating Peierls pseudogap while the structure in $\sigma(\omega)$ at 40 K is due to transitions across the real Peierls gap $2\Delta(0) \cong 0.2\,\mathrm{eV}$ (Fig. 5.16). Note that the resonance at the gap frequency $\omega_g \cong 16000\,\mathrm{cm}^{-1}$ is the associated longitudinal excitation. The latter can be excited if p-polarized light under non-normal incidence is reflected from a crystal surface which is oriented perpendicular to the z-axis as has been shown by *Brüesch* [5.42] and by *Agroskin* et al. [5.43].

The temperature dependence of the reflectivity in the FIR for light polarized parallel to the conducting axis, as observed by *Brüesch* [5.44] and by *Brüesch* et al. [5.40], is shown in Fig. 5.17a. The very strong structure observed at 4.2 K becomes weaker and smeared out with increasing temperature and disappears around 200 K. This is due to the temperature dependence of the damping of the mode as well as to screening effects caused by thermally activated quasi-free electrons. Figure 5.17b shows the corresponding behaviour of the frequency dependent conductivity: $\sigma(\omega, T)$ has been obtained from an oscillator fit with a frequency-dependent damping

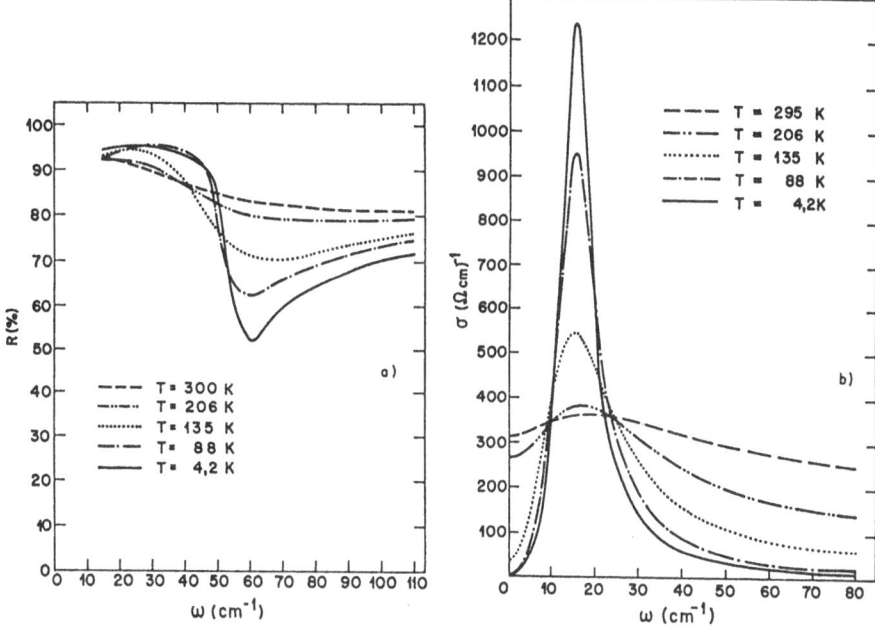

Fig. 5.17. (a) Temperature dependence of the reflectivity observed in the far-infrared for light polarized parallel to the conducting axis. A strong structure with a very high oscillator strength is observed at 4.2 K which is assigned to the pinned Fröhlich mode. The increasing damping smears out the structure with increasing temperature. **(b)** Temperature dependence of the optical conductivity σ in the far-infrared. The dc conductivity for $T \geq 135$ K is mainly due to single particle contributions which has been added to the oscillator term of the pinned Fröhlich mode [5.40]

to the reflectivity data shown in Fig. 5.17a. The details of the model used for the fit have been discussed in [5.40]. The optical conductivity of KCP as determined by a Kramers-Kronig analysis of the reflectivity data at 40 K is shown in Fig. 5.18. Note the similaritiy of this figure with the schematic Fig. 5.6b. There are two peaks in the optical conductivity $\sigma(\omega)$. The very strong peak centered around $1600\,\mathrm{cm}^{-1}$ is due to transitions across the Peierls gap. The low-frequency mode centered around $15\,\mathrm{cm}^{-1}$ is assigned to the pinned Fröhlich mode, i.e. to the phase-modulating A_- mode of the CDW (Fig. 5.5b), and in the following the reasons for this assignment will be discussed [5.39, 40].

At $T = 0$ the response of the pinned CDW can be discussed phenomenologically in terms of an oscillator model

$$\varepsilon = \varepsilon_\mathrm{g} + \varepsilon_\mathrm{g}(\omega_\mathrm{LO}^2 - \omega_\mathrm{TO}^2)/(\omega_\mathrm{TO}^2 - \omega^2 - \mathrm{i}\Gamma\omega) \ , \tag{5.67}$$

where ε_g is the "high-frequency" dielectric constant due to transitions across the Peierls gap, ω_TO the oscillator frequency defined by the pinning force consant, compare (5.62), and Γ a damping term. ω_LO, the plasma frequency

133

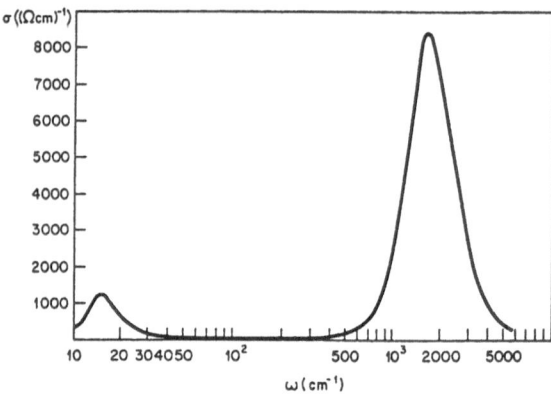

Fig. 5.18. Optical conductivity of KCP as determined by a Kramers-Kronig analysis of the reflectivity at 40 K. The very strong resonance at 1600 cm^{-1} is due to transitions across the Peierls gap while the small peak centered around 15 cm^{-1} is due to the pinned Fröhlich mode. Note the logarithmic frequency scale

of the CDW, is related to the effective mass m^*_{CDW} of the collective excitation by

$$\omega^2_{\text{LO}} = \frac{4\pi ne^2}{m^*_{\text{CDW}}\varepsilon_g} \;, \tag{5.68}$$

where n defines the density of the condensed electrons. It turns out that a fit to the 4.2 K data of Fig. 5.17a is rather insensitive to the value of ω_{TO}. The frequency ω_{TO} has therefore been determined by using the known value of the static dielectric constant $\varepsilon_0 \cong 3000$ obtained from microwave data [5.45]. By this procedure one obtains $\omega_{\text{LO}} = 58\,\text{cm}^{-1}$, $\omega_{\text{TO}} = 15\,\text{cm}^{-1}$, $\varepsilon_g = 190$ and $\Gamma = 7.5\,\text{cm}^{-1}$ at 4.2 K. Assuming all electrons to be condensed in the CDW, an effective mass $m^*_{\text{CDW}} = 980$ is deduced from (5.68).

Mean-field theory allows the determination of the electron-phonon coupling constant λ in two independent ways: from (5.58) with $2\Delta(0) = 0.2\,\text{eV}$, $E_F = 3.25\,\text{eV}$, one obtains $\lambda \cong 0.2$. Independently, λ is given by (5.63); with $m^*_{\text{CDW}}/m_e \cong 10^3$ and $\omega_{2k_F} \cong 10\,\text{meV}$ as the unrenormalized phonon frequency, one obtains $\lambda \cong 0.4$. The discrepancy between the two values is partly due to the assumption of an energy-independent electron-phonon interaction.

Using (5.59) with $2\Delta(0) \cong 0.2\,\text{eV}$ one obtains $T^{\text{MF}}_{\text{P}} \cong 600\text{–}800\,\text{K}$; a similar value is obtained from (5.54) with $E_F = 3.25\,\text{eV}$ and $\lambda = 0.2$. This implies that a fluctuating Peierls gap with a mean gap energy of about $0.2\,\text{eV}$ has already established at 300 K, as discussed in Sect. 5.5.3.

The pinned phase mode arises essentially from the high-temperature $(\pi/a, \pi/a, 2k_F)$ phonons (a: distance between neighbouring chains) as shown in Fig. 5.13. From diffuse X-ray experiments it follows that below T^{MF}_{P} a superstructure develops which transforms the volume of the high-temperature unit cell $a \times a \times 2c$ (c: Pt-Pt distance along the chains) into a low temperature unit cell of dimension $2a \times 2a \times (\sim 6c)$. The associated reduction of the Brillouin zone transforms the $(\pi/a, \pi/a, \pm 2k_F)$ phonons into a $q = (0, 0, 0)$ zone center optical phonon, which can, in principle, absorb light, if it induces a dipole moment. The non-vanishing dipole moment, however, is a consequence of the fact that the Peierls distortion induces a CDW. Since the Peierls distortion of an adjacent Pt chain exhibits a phase change of π [5.29], the CDW also exhibits such a phase change in adjacent chains for the $(\pi/a, \pi/a, 2k_F)$ mode as shown in Fig. 5.19 [5.39]. The arrows in Fig. 5.19 indicate the vibrational mode of the Pt ions about the displaced mean equilibrium positions; it is obvious that this mode induces a dipole moment since opposite charges $\pm e^*$ vibrate against each other similar to the $q = 0$ TO mode in NaCl.

It should be pointed out that no other interpretation of the observed FIR structure than that in terms of a pinned CDW is possible. The phenomenological equation (5.67) not only describes bound electronic systems but also lattice vibrations and coupled electron-phonon systems such as the Peierls CDW. From the effective mass $m^*_{CDW} \cong 10^3 \, m_e$, all lattice vibrations (except possibly proton motion) and also purely electronic excitations can be excluded. Protons are present in the system but at low symmetry sites [5.16]. If they were responsible for the structure, they would exhibit a component in the transverse direction, which is not observed. Furthermore, several studies on deuterated samples show no change in the relevant

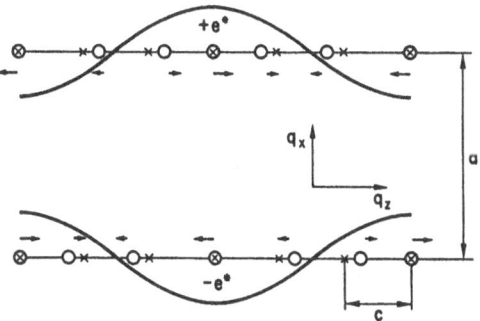

Fig. 5.19. The figure shows the phase modulating A_- mode considered by Fröhlich in the Peierls distorted phase. The mode originates from the high temperature $(\pi/a, \pi/a, \pm 2k_F)$ phonons which are converted into $q = 0$ phonons in the low temperature phase with unit volume $2a \times 2a \times (\sim 6c)$. $\times$ marks the equidistant equilibrium positions at $T > T_P^{MF}$ and ($\circ$) marks the equilibrium positions at $T \ll T_P^{MF}$ in the Peierls distorted phase. ($-$) shows the induced CDW with amplitude e^* which is equal but opposite in phase in adjacent Pt chains. Based on the z-polarized vibration as indicated by the arrows a longitudinal (LO) mode propagating in the z direction and a transverse (TO) mode propagating in the (x, y) direction exist. The A_- mode is infrared active but not Raman active [5.39, 44]

135

physical properties. Thus the effective mass or the huge associated oscillator strength can only be caused by a coupled electron-phonon system, which in our case is the CDW.

The question arises as to which extend the dynamics of the CDW contributes to the observed dc conductivity. In a 1-DC the dc conductivity has two contributions

$$\sigma_{dc} = \sigma_{sp} + \sigma_{CDW} \, . \tag{5.69}$$

Here, σ_{sp} is the single particle contribution which for quasi-free electrons is given by

$$\sigma_{sp} = \frac{ne^2}{m^* \Gamma_{sp}} \, , \tag{5.70}$$

where n is the electron concentration, m^* the effective mass and Γ_{sp} the damping. σ_{CDW} is the contribution of the CDW. As we have already mentioned at the end of Sect. 5.2, the CDW will oscillate about its pinned positions at low temperature, but with increasing temperature its amplitude will increase and reach such high values around the pinning temperature $T_P(\cong 150\,\mathrm{K}$ in KCP), that the CDW may hop a certain distance and carry current similar to the hopping of oscillating ions in an ionic conductor. At room temperature the contribution of the CDW is

$$\sigma_{CDW} = \frac{ne^2}{m^*_{CDW} \Gamma_{CDW}} \, , \tag{5.71}$$

where m^*_{CDW} is the effective mass and Γ_{CDW} the damping of the CDW. Due to the high effective mass of the CDW, $m^*_{CDW} \cong 10^3 \, m_e$, it is estimated that the contribution of the CDW to σ_{dc} is less than 50% [5.25, 40].

The amplitude mode is shown schematically in Fig. 5.20 [5.44]. Inspection of this mode shows that it modulates the zz-component of the

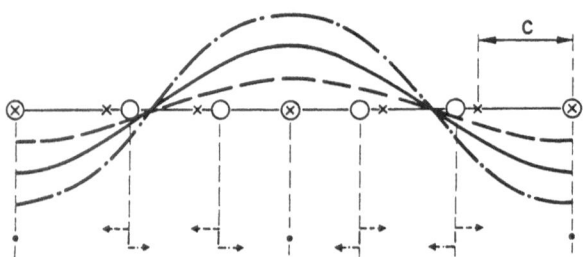

Fig. 5.20. Amplitude modulating A_+ mode for KCP. The Peierls distortion leads to a superstructure with period $\sim 6c$ (c: Pt-Pt distance). (x) mark the equidistant equilibrium positions of the Pt ions above T_P^{MF}. ($-$) is the CDW for the static distortion. The Pt ions vibrate around their displaced equilibrium positions as shown by the arrows. This modulates the amplitude of the CDW as shown by ($- - -$). The nodes of the CDW remain fixed. The A_+ mode is Raman active but not infrared active [5.44]

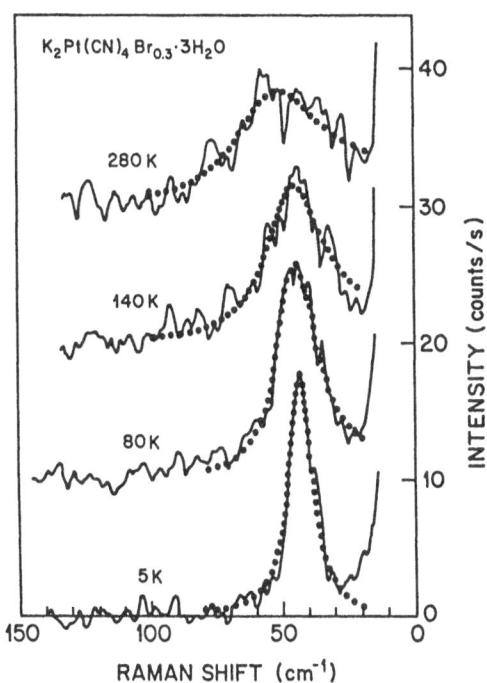

Fig. 5.21. Raman spectrum of KCP in the $x(zz)\bar{x}$ configuration showing the CDW amplitude mode at various temperatures. The curves are displaced by 10 counts/s each. (• • •) represents the one-oscillator fit to the experimental results. Note that $x(yy)\bar{x}$ does not show this mode [5.46]

electronic polarizability and should therefore be Raman active. *Steigmeier* et al. [5.46] have observed a Raman line of A_1 symmetry (space group D_{4h}) at 44 cm^{-1} which shows interesting temperature dependent properties as shown in Fig. 5.21. From the observed longitudinal optical frequency $\omega_{LO} = 58$ cm^{-1} associated with the pinned phase mode and using (5.65) the frequency of the A_1 mode is expected to be 47 cm^{-1} which is in excellent agreement with the observed frequency. From this agreement in frequency as well as from a comparison of symmetry and scattering intensity of this line with theoretical predictions it is concluded that the excitation concerned represents the amplitude mode of the CDW. Both, the pinned phase mode and the amplitude mode are roughly independent on temperature below 300 K which is consistent with $T_P^{MF} \cong 600$–800 K.

5.6 Effects of Fluctuations and Three-Dimensional Coupling

It has been well established that due to thermal fluctuations, one-dimensional systems without long-range forces cannot have a phase transition at $T > 0$ [5.4, 47, 48]. The electron-phonon interaction does not come into the category of long-range forces and thus there can be no Peierls transition in strictly one-dimensional systems. Therefore the mean field theory of the Peierls instability which was developed in Sects. 5.3, 4 for a one-dimensional system cannot give an adequate description of the observed phenomena. The reason why a phase transition was obtained for a one-dimensional system by the mean field approximation is very simple: Fluctuations have been suppressed in that all lattice modes other than those with $2k_F$ wave vector dependence were ignored. Therefore, the short-range electron phonon interaction was effectively turned into a long-range, sinusoidal force of wavevector $2k_F$. If all the Fourier components were included, the short-range character of the force would be restored and the phase transition suppressed [5.7, 49].

The only way in which it is possible to obtain a phase transition at $T>0$ for the electron-phonon system is to take into account three-dimensional coupling. A weak interchain coupling must suppress partially the effects of thermal fluctuations since the static Peierls transition is observed to occur at a finite temperature T_P that is considerably lower than the expected mean field transition temperature T_P^{MF}. Fortunately the development of the 1-D mean field approximation has not been a vaste of time as it displays qualitatively many of the features which are important when recourse is made to the three-dimensional case. In a more adequate theory, however, the thermal fluctuations which suppress the Peierls transition must be included, and three-dimensional coupling must then be incorporated in order to obtain a Peierls transition at $T>0$. A large number of theoretical papers have addressed themselves to this extremely interesting interplay between fluctuation and order phenomena [5.20, 50–53].

Physically one has to distinguish between two temperatures, a temperature T_P^{Ph} below which a CDW gets established and a temperature $T_P = T_P^{3D}$ at which long range order sets in and the CDW gets pinned [5.5, 54]. In other words, the two relevant parameters are $\sqrt{\langle Q_{2k_F}^2 \rangle}$ and $\langle Q_{2k_F} \rangle$, where Q_{2k_F} is the amplitude of the CDW. In the true metallic region above T_P^{MF} which for KCP is not experimentally accessible, $\sqrt{\langle Q_{2k_F}^2 \rangle}$ and $\langle Q_{2k_F} \rangle = 0$. The intermediate region with $\sqrt{\langle Q_{2k_F}^2 \rangle} \neq 0$ and $\langle Q_{2k_F} \rangle = 0$ is still metallic in the sense that there is no true energy gap but is governed by fluctuation effects which strongly depress the dc conductivity and cause a central peak in the inelastic neutron data; the relevant temperature is the "phonon-dominated" temperature T_P^{Ph}. At low temperatures (below about 100 K for KCP) $\langle Q_{2k_F} \rangle \neq 0$, long range order sets in and below T_P^{3D} the CDW is pinned due to impurities and three-dimensional coupling, and semiconductor properties are found; pinning due to 3-D coupling starts at the temperature T_P^{3D} which for the case of weak interchain coupling is given by (5.72). The "phase diagram" of a quasi 1-DC is schematically illustrated in Fig. 5.22.

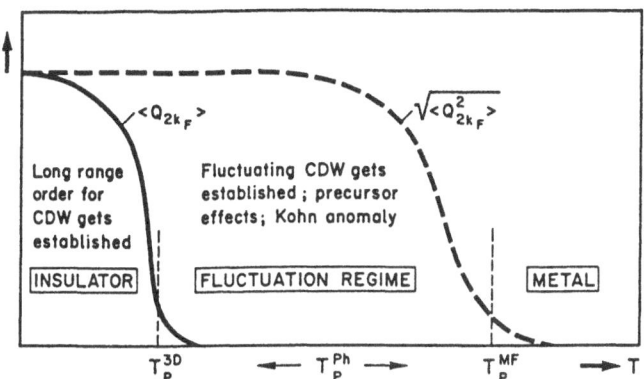

Fig. 5.22. Schematic illustration of the "phase diagram" of a quasi one-dimensional conductor

According to the theory of *Rice* and *Straessler* [5.41] the combined effects of fluctuation and weak three-dimensional coupling leads to the following expression for the temperature at which the CDW gets pinned:

$$T_P^{3D} = 0.28\,(c_\perp/\lambda)^{1/4}T_P^{MF} \ . \tag{5.72}$$

Here, the coupling constant $c_\perp$ determines the degree of interchain coupling and λ is the electron-phonon coupling constant given by (5.55). For vanishing 3-D coupling $(c_\perp = 0)$, $T_P^{3D} = 0$ and a static Peierls distortion will not be established. For KCP the relevant temperatures are estimated to be $T_P^{MF} \cong 600\,\mathrm{K}$, $T_P^{Ph} \cong 200\,\mathrm{K}$ and $T_P^{3D} \cong 100\,\mathrm{K}$.

The picture is greatly simplified by the fact that in the experimentally accessible temperature region of KCP the rms value of the amplitude of the CDW, $\sqrt{\langle Q_{2k_F}^2 \rangle}$, has already essentially reached its zero-temperature value. As a consequence the optically observed pseudogap will occur at the zero temperature gap value. Even at room temperature where fluctuations causes a strong depression of the low-frequency conductivity, this occurs in the energy region of the zero-temperature gap (Fig. 5.16),

The work on KCP has clearly demonstrated the importance of two typical concepts of 1-D physics, namely the Peierls instability and the effects of fluctuations. KCP is sufficiently one-dimensional to exhibit a Peierls instability, and in an extremely wide temperature region the physical properties are governed by fluctuations. Residual three-dimensionality is responsible for the pinning of the CDW below about 100 K.

6. Phonons in Disordered Systems
(By J. Bernasconi)

In this chapter we study the effects of defects and disorder on phonons. In Sect. 6.2 we introduce the Green's function formalism, an important tool for the investigation of disordered systems. The linear chain with isolated defects is analyzed in Sect. 6.3, and in Sect. 6.4 the random binary mass chain is studied within the coherent potential approximation (CPA). We also describe cluster extensions of the CPA and a recently developed renormalization-group approach, and compare the predictions of the different approximations with numerical results. In Sect. 6.5 we introduce the T-matrix formalism and indicate how it is used to apply the CPA to three-dimensional lattices and to more complicated types of disorder. In Sect. 6.6, finally, some selected experimental results are compared with the predictions of theoretical calculations.

6.1 The Effects of Defects and Disorder on Phonons

In a perfect crystalline material, the translational invariance of the lattice considerably simplifies a theoretical investigation of the vibrational properties. Interesting real materials, however, are often disordered (e.g., alloys), and even carefully prepared crystals invariably contain defects or impurities. The corresponding perturbation, or even destruction, of the translational and point symmetries of the crystal modifies both the vibrational eigenfrequencies and eigenvectors of the normal modes. If the perturbation by a defect is large enough, modes can be driven out of the host crystal continuum ($\rightarrow$ localized modes), or resonances may occur within the continuum ($\rightarrow$ quasi localized modes). In strongly disordered systems (large defect concentrations) new phenomena are observed, such as localized states due to clusters of defects, or localization of states due to the absence of long-range order (Anderson localization).

The effects of defects and disorder on one-phonon spectra are most conveniently studied via infrared absorption, Raman scattering, and neutron scattering experiments. Except for a few special cases, however, the corresponding theoretical problems cannot be solved exactly. Numerical, as well as a number of approximate analytical methods have been developed to investigate the vibrational properties of disordered systems. Most analytical methods are concerned with the calculation of ensemble (or configuration) averaged quantities, and numerical results provide important tests for the validity of such analytical approximations.

In this chapter we shall, of course, not attempt to give a thorough review on the vibrational properties of disordered materials, or to present a detailed analysis of the various theoretical approaches[1]. Instead, we shall restrict ourselves to a brief description and illustration of some important theoretical methods by applying them to simple linear chain systems. The corresponding treatment of more complicated disordered systems is discussed on a rather general level only, and a comparison between theoretical calculations and experimental results is restricted to a few illustrative cases.

6.2 Green Functions

The equations of motion for a harmonic linear chain of masses m_i and nearest-neighbour force constants $f_{i,i+1}$ are given by

$$m_i \ddot{u}_i = f_{i-1,i}(u_{i-1} - u_i) + f_{i,i+1}(u_{i+1} - u_i) + F_i \; , \tag{6.1}$$

where u_i denotes the displacement of mass m_i, and F_i the force acting on m_i.

If we consider the special case of a short (δ-function like) unit force pulse applied at time $t = 0$ to a single mass m_{i_0}, the corresponding displacements are called *Green functions* of the problem, and we shall denote them by $g_{ii_0}(t)$:

$$m_i \ddot{g}_{ii_0} = f_{i-1,i}(g_{i-1,i_0} - g_{ii_0})$$
$$+ f_{i,i+1}(g_{i+1,i_0} - g_{ii_0}) + \delta_{ii_0}\delta(t) \; . \tag{6.2}$$

As initial conditions we assume that

$$g_{ik}(t = 0) = \dot{p}_{ik}(t = 0) = 0 \; , \tag{6.3}$$

and introduce the Laplace-transformed Green functions $G_{ik}(z) = \tilde{g}_{ik}(z^{1/2})$, where

$$\tilde{g}_{ik}(z) = \int_0^\infty dt \, e^{-zt} g_{ik}(t) \; . \tag{6.4}$$

---

[1] The reader is referred to the excellent reviews listed in [6.1–9].

Equations (6.2 and 3) then lead to

$$m_i z G_{ii_0} = f_{i-1,i}(G_{i-1,i_0} - G_{ii_0})$$
$$+ f_{i,i+1}(G_{i+1,i_0} - G_{ii_0}) + \delta_{ii_0} \ , \tag{6.5}$$

and corresponding Green functions may be defined in a completely analogous way for two- or three-dimensional lattices. The Green function $g_{ik}(t)$, or $G_{ik}(z)$, thus describes the response of mass m_i due to a δ-function-like stimulation of mass m_k.

Green functions have become an important tool for the calculation of the vibrational density of states and of thermally averaged observable quantities in disordered systems [6.5–9]. The phonon density of states of our linear chain, e.g.,

$$n(\omega) = 2\omega N(\omega^2) \ , \tag{6.6}$$

can be expressed in terms of the $G_{ik}(z)$ as

$$N(\omega^2) = \frac{1}{\pi N} \mathrm{Im}\left\{ \mathrm{Tr}\{m_i^{1/2} G_{ik}(-\omega^2 - i0^+) m_k^{1/2}\} \right\}$$
$$= \frac{1}{\pi N} \sum_i m_i \mathrm{Im}\{G_{ii}(-\omega^2 - i0^+)\} \ , \tag{6.7}$$

where $N(\to \infty)$ denotes the total numbers of masses in the chain, and 0^+ an infinitesimally small positive quantity. The local density of states for the response of mass m_i, due to an instantaneous stimulation there,

$$N_i(\omega^2) = \frac{1}{\pi} m_i \mathrm{Im}\{G_{ii}(-\omega^2 - i0^+)\} \ , \tag{6.8}$$

is thus directly given by the imaginary part of the diagonal Green function, $\mathrm{Im}\{G_{ii}(z)\}$. The stimulus may be provided by different means, and a number of thermally averaged observable quantities are therefore also determined by the $\mathrm{Im}\{G_{ik}(z)\}$. One can show, e.g., that the one-phonon infrared absorption and Raman scattering coefficients are both proportional to an expression of the type [6.1, 5, 7–9]

$$\sum_{i,k} q_i q_k \mathrm{Im}\{G_{ik}(-\omega^2 - i0^+)\} \ .$$

To establish such relations one often uses the double-time thermodynamic Green functions [6.10], instead of the so-called classical Green functions defined by (6.5), because the former are directly connected with thermally averaged correlation functions. In the lattice vibration problem, however, both type of Green functions satisfy the same equations, so that for our purposes we can restrict ourselves to the classical ones introduced above. Thermally averaged correlation functions have then to be calculated via an extension of Nyquist's theorem. As an example, the mean square displacement of atom i, $\langle u_i(t)^2 \rangle$, can be expressed as [6.7]

$$\langle u_i(t)^2 \rangle = \frac{\hbar}{\pi} \int_0^\infty d\omega \, \mathrm{ctgh}\left(\frac{1}{2}\beta\omega\right) \mathrm{Im}\{G_{ii}(-\omega^2 - i0^+)\} \ , \tag{6.9}$$

where $\beta = 1/k_B T$ is the inverse temperature.

In disordered systems one is usually interested in configuration averaged quantities, and different methods to evaluate configuration averaged Green functions have been developed (Sects. 6.3 and 4). In linear chain systems, we can often take advantage of the fact that the equations of motion for the $G_{ik}(z)$, (6.5), can be solved formally in terms of infinite continued fractions [6.11, 12]:

To simplify the notation, we assume without loss of generality that the stimulation occurs at site $i_0 = 0$ and then denote $G_{i0}(z)$ by $G_i(z)$. Equations (6.5) then become

$$m_i z G_i = f_{i-1,i}(G_{i-1} - G_i) + f_{i,i+1}(G_{i+1} - G_i) + \delta_{i0} \ , \tag{6.10}$$

and we introduce the quantities

$$X_i = f_{i,i+1}\frac{G_i - G_{i+1}}{G_i} \ , \quad i = 0, 1, \ldots \ , \tag{6.11a}$$

and

$$Y_i = f_{-i-1,-i}\frac{G_{-i} - G_{-i-1}}{G_{-i}} \ , \quad i = 0, 1, \ldots \ . \tag{6.11b}$$

The X_i and Y_i can then be expressed as infinite continued fractions depending on z and on the m_i and $f_{i,i+1}$. They are recursively determined by

$$X_i = \cfrac{1}{\cfrac{1}{f_{i,i+1}} + \cfrac{1}{m_{i+1}z + X_{i+1}}} \quad \text{and} \tag{6.12a}$$

$$Y_i = \cfrac{1}{\cfrac{1}{f_{-i-1,-i}} + \cfrac{1}{m_{-i-1}z + Y_{i+1}}} \ , \tag{6.12b}$$

respectively, and the formal solution for the $G_i(z)$ becomes

$$G_0 = \frac{1}{m_0 z + X_0 + Y_0} \ , \tag{6.13a}$$

$$G_i = G_0 \prod_{k=1}^{i} \frac{X_{k-1}}{m_k z + X_k} \quad , \quad i = 1, 2, \dots , \tag{6.13b}$$

$$G_{-i} = G_0 \prod_{k=1}^{i} \frac{Y_{k-1}}{m_{-k} z + Y_k} \quad , \quad i = 1, 2, \dots . \tag{6.13c}$$

The above formal expressions turn out to be very useful for the evaluation of Green functions in linear chains with defects (Sect. 6.3), or for the approximate calculation of averaged Green functions in randomly disordered chains (Sect. 6.4).

Before we turn to disordered systems, however, let us first calculate the diagonal Green function, $G_0(z)$, and the phonon density of states for a perfect monatomic and diatomic linear chain, respectively.

Consider a monatomic linear chain with all masses m_i equal to m and all force constants $f_{i,i+1}$ equal to f. It follows that $Y_i = X_i = X$, independent of i, and we have

$$G_0 = \frac{1}{mz + 2X} \, , \tag{6.14}$$

$$X = \frac{1}{1/f + (1/(mz + X))} \, . \tag{6.15}$$

Equation (6.15) is a quadratic equation for X, and the solution is given by[2]

$$X = \frac{m}{2} [z^{1/2} (\omega_m^2 + z)^{1/2} - z] \, , \quad \text{where} \tag{6.16}$$

$$\omega_m^2 \equiv 4 \frac{f}{m} \, , \quad \text{so that} \tag{6.17}$$

$$G_0 = \frac{1}{mz^{1/2}(\omega_m^2 + z)^{1/2}} \, . \tag{6.18}$$

The density of states

$$N(\omega^2) = \frac{1}{\pi} \text{Im}\{m G_0(-\omega^2 - i0^+)\} \, , \tag{6.19}$$

then becomes

---

[2] Note that we have to write $z^{1/2}(\omega_m^2 + z)^{1/2}$, and not $[z(\omega_m^2 + z)]^{1/2}$, to ensure that $\tilde{g}_0(z) = G_0(z^2)$ possesses the correct analyticity properties of a Laplace transform.

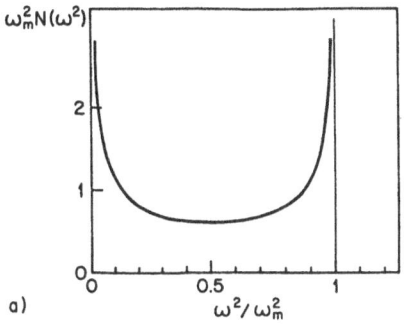
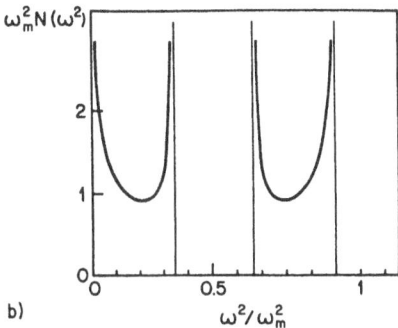

Fig. 6.1a,b. Density of states of a monoatomic chain (a), and of a diatomic chain with $m_1 = 2m_2$ (b)

$$N(\omega^2) = \begin{cases} \dfrac{1}{\pi[\omega^2(\omega_m^2 - \omega^2)]^{1/2}} \;, & 0 \leq \omega \leq \omega_m \\ 0 \;, & \omega > \omega_m \;, \end{cases} \tag{6.20}$$

a well-known result which is displayed in Fig. 6.1a.

The diatomic chain is a little more complicated. We have to calculate two diagonal Green functions, $G_0^{(1)}$ and $G_0^{(2)}$, corresponding to the stimulation of an m_1-mass and an m_2-mass, respectively. If all force constants are equal to f, we have

$$G_0^{(1)} = \frac{1}{m_1 z + 2X_1} \;, \tag{6.21a}$$

$$G_0^{(2)} = \frac{1}{m_2 z + 2X_2} \;, \quad \text{and} \tag{6.21b}$$

$$X_1 = \frac{1}{1/f + 1/(m_2 z + X_2))} \;, \tag{6.22a}$$

$$X_2 = \frac{1}{1/f + (1/(m_1 z + X_1))} \;. \tag{6.22b}$$

The determination of X_1 and X_2 involves the solution of a quadratic equation, and a straightforward calculation then leads to

$$m_1 G_0^{(1)}(z) = \frac{(\omega_2^2 + z)^{1/2}}{z^{1/2}(\omega_1^2 + z)^{1/2}(\omega_m^2 + z)^{1/2}} \tag{6.23a}$$

$$m_2 G_0^{(2)}(z) = \frac{(\omega_1^2 + z)^{1/2}}{z^{1/2}(\omega_2^2 + z)^{1/2}(\omega_m^2 + z)^{1/2}} \;, \tag{6.23b}$$

where

$$\omega_1^2 \equiv \frac{2f}{m_1} \quad , \quad \omega_2^2 \equiv \frac{2f}{m_2} \quad , \quad \omega_m^2 \equiv \omega_1^2 + \omega_2^2 \ . \tag{6.24}$$

For $m_1 > m_2$, i.e. $\omega_1 < \omega_2 < \omega_m$, the density of states,

$$N(\omega^2) = \frac{1}{2\pi} \left[\mathrm{Im}\{m_1 G_0^{(1)}(z)\} + \mathrm{Im}\{m_2 G_0^{(2)}(z)\} \right]_{z=-\omega^2-i0^+} \quad , \tag{6.25}$$

becomes

$$N(\omega^2) = \begin{cases} \dfrac{1}{2\pi} \dfrac{\omega_m^2 - 2\omega^2}{[\omega^2(\omega_1^2 - \omega^2)(\omega_2^2 - \omega^2)(\omega_m^2 - \omega^2)^{1/2}]} & 0 \leq \omega \leq \omega_1 \\[2mm] 0 & \omega_1 < \omega < \omega_2 \\[2mm] \dfrac{1}{2\pi} \dfrac{2\omega^2 - \omega_m^2}{[\omega^2(\omega^2 - \omega_1^2)(\omega^2 - \omega_2^2)(\omega_m^2 - \omega^2)^{1/2}]} & \omega_2 \leq \omega \leq \omega_m \\[2mm] 0 & \omega > \omega_m \end{cases} \tag{6.26}$$

and thus consists of two distinct branches, an accoustical one ($0 \leq \omega \leq \omega_1$) and an optical one ($\omega_2 \leq \omega \leq \omega_m$), see Fig. 6.1b.

6.3 The Linear Chain with Isolated Defects

In the limit of very small defect concentrations it turns out to be sufficient to consider a single defect in an otherwise perfect crystal. This is obvious for experiments where the defect itself is used as a probe, e.g. for the one-phonon optical absorption due to charged defects in a homopolar crystal. Such an experiment measures directly the frequency response of the defects, and for small defect concentrations c, the absorption coefficient is simply given by [6.1]

$$\alpha(\omega) \propto c\omega \, \mathrm{Im} \left\{ G_0^{(d)}(-\omega^2 - i0^+) \right\} \ , \tag{6.27}$$

were $G_0^{(d)}(z)$ is the diagonal Green function at the defect for the correspond-ing single defect problem.

Other observable quantities depend on the phonon spectrum of the per-turbed lattice, and here it is less obvious that for small c the corresponding calculations can be reduced to a single defect problem. It can be shown, however, that the coherent potential approximation (which is a single de-fect theory, see Sect. 6.4.2) gives an answer correct to order c. Moreover, this single defect theory often represents a reasonably accurate approximation also for non-dilute disordered systems.

Single defect problems can be solved exactly, and in the following we shall analyze in some detail the effects of a single mass defect and of a double force constant defect, respectively, in a monatomic linear chain.

6.3.1 The Single Mass Defect

Consider a single defect of mass m_0 in a monatomic linear chain of masses m and force constants f. The position of the defect is assumed to be the origin, and from (6.13 and 12) the diagonal Green function for the defect, $G_0(z)$, is then seen to be given by

$$G_0(z) = \frac{1}{m_0 z + 2X(z)} \ , \quad \text{where} \tag{6.28}$$

$$X = \frac{1}{1/f + (1/(mz + X))} \ . \tag{6.29}$$

Equation (6.29) is easily solved, and we obtain

$$X(z) = \frac{m}{2}[z^{1/2}(\omega_m^2 + z)^{1/2} - z] \ , \tag{6.30}$$

and finally

$$G_0(z) = \frac{1}{m[z^{1/2}(\omega_m^2 + z)^{1/2} - \varepsilon z]} \ , \quad \text{with} \tag{6.31}$$

$$\omega_m^2 \equiv 4\frac{f}{m} \ , \quad \text{and} \tag{6.32}$$

$$\varepsilon \equiv \frac{m - m_0}{m} \ . \tag{6.33}$$

The corresponding local density of states contribution,

$$N_0(\omega^2) = \frac{m_0}{\pi}\text{Im}\{G_0(-\omega^2 - i0^+)\} \ , \tag{6.34}$$

as well as the real part of $G_0(-\omega^2 - i0^+)$ are plotted in Fig. 6.2 for different values of the relative mass defect ε. We note that a localized mode, corresponding to a resonance of $G_0(-\omega^2 - i0^+)$, only occurs if $\varepsilon > 0$ (i.e., $m_0 < m$). In this case, the frequency of this localized mode, ω_{loc}, is given by

$$\frac{\omega_{\text{loc}}^2}{\omega_m^2} = \frac{1}{1 - \varepsilon^2} \ , \quad 0 < \varepsilon < 1 \ . \tag{6.35}$$

The mean square amplitude, $|\chi_0(\omega)|^2$, of the defect atom in a normal mode of frequency ω is given by

$$|\chi_0(\omega)|^2 = \frac{1}{\pi}\text{Im}\{G_0(-\omega^2 - i0^+)\} \ . \tag{6.36}$$

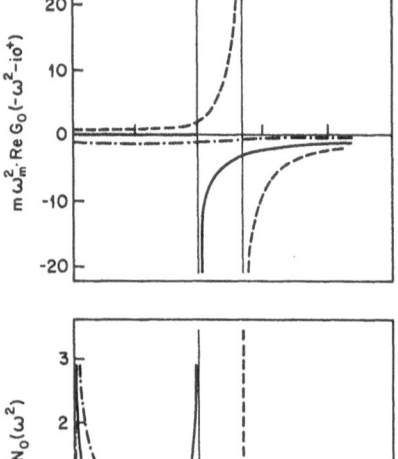

Fig. 6.2. Behavior of $m\mathrm{Re}\{G_0(-\omega^2 - i0^+)\}$ and of $N_0(\omega^2) = (m_0/\pi)\,\mathrm{Im}\{G_0(-\omega^2 - i0^+)\}$ for a monatomic linear chain with a single mass defect. —— $m_0 = m$ ($\varepsilon = 0$, pure chain); ---- $m_0 = m/2$ ($\varepsilon = 1/2$); ·–·–·– $m_0 = 2m$ ($\varepsilon = -1$)

For the localized mode this leads to

$$m_0|\chi_0(\omega_{\mathrm{loc}})|^2 = \frac{2\varepsilon}{1+\varepsilon}\ , \qquad 0<\varepsilon<1 \ , \tag{6.37}$$

and for the band normal modes the mean square amplitude of the defect atom, relative to that in the perfect chain, becomes

$$\frac{|\chi_0(\omega)|^2}{|\chi_0(\omega)|^2_{\varepsilon=0}} = \left(1 + \varepsilon^2\,\frac{\omega^2/\omega_m^2}{1 - \omega^2/\omega_m^2}\right)^{-1}\ , \qquad 0<\omega<\omega_m \ . \tag{6.38}$$

In our linear chain system, we thus observe no resonances within the band of normal modes. This is in contrast to three-dimensional lattices, where a heavy defect usually leads to a resonance. For a double force constant defect, however, such resonances also occur in a linear chain system, as we shall see in the following section.

6.3.2 The Double Force Constant Defect

Let us now consider a monatomic chain with a double force constant defect. The defect atom is assumed to have the same mass, m, as the host atoms, but the two neighboring force constants are changed from f to f_0 (Fig. 6.3). According to (6.13 and 12), the diagonal Green function for the defect atom, $G_0(z)$, is then given by

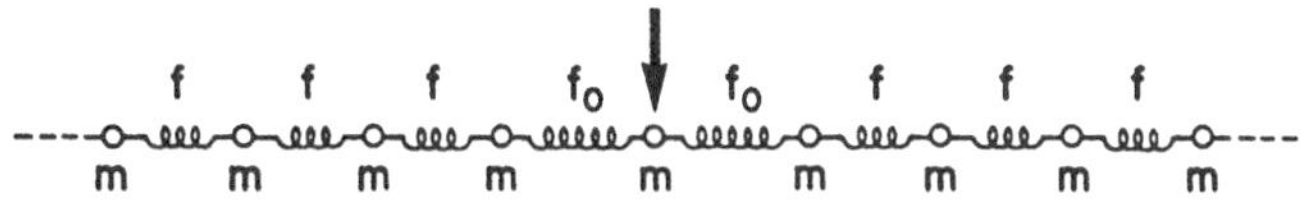

Fig. 6.3. Monatomic linear chain with a double force constant defect

$$G_0 = \frac{1}{mz + 2X_0} \; , \tag{6.39}$$

$$X_0 = \frac{1}{1/f_0 + (1/(mz + X))} \; , \tag{6.40}$$

$$X = \frac{1}{1/f + (1/(mz + X))} \; . \tag{6.41}$$

This leads to the following explicit expression for $G_0(z)$

$$G_0(z) = \frac{1}{m\left(z + \dfrac{(1-\eta)\omega_m^2\left[z^{1/2}(\omega_m^2+z)^{1/2}+z\right]}{(1-\eta)\omega_m^2+2\left[z^{1/2}(\omega_m^2+z)^{1/2}+z\right]}\right)} \; , \tag{6.42}$$

with

$$\omega_m^2 \equiv 4\frac{f}{m} \; , \quad \text{and} \tag{6.43}$$

$$\eta \equiv \frac{f - f_0}{f} \; . \tag{6.44}$$

Figure 6.4 exhibits the behavior of the corresponding local density of states contribution,

$$N_0(\omega^2) = \frac{m}{\pi}\text{Im}\{G_0(-\omega^2 - i0^+)\} \; , \tag{6.45}$$

and of the real part of $G_0(-\omega^2 - i0^+)$ for different values of the relative force constant defect η. We see that a double force constant defect leads to a localized mode if $\eta < 0$ (i.e., $f_0 > f$). Its frequency, ω_{loc}, is calculated to be given by

$$\frac{\omega_{\text{loc}}^2}{\omega_m^2} = \frac{1 - \eta}{8\eta}\{1 + 3\eta - [(1 - \eta)(1 - 9\eta)]^{1/2}\} \; , \quad \eta < 0 \; . \tag{6.46}$$

If $f_0 < f \; (\eta > 0)$, on the other hand, we observe resonances within the band of normal modes. These resonances are most conveniently studied by considering the mean square amplitude of the defect atom relative to that in the perfect chain,

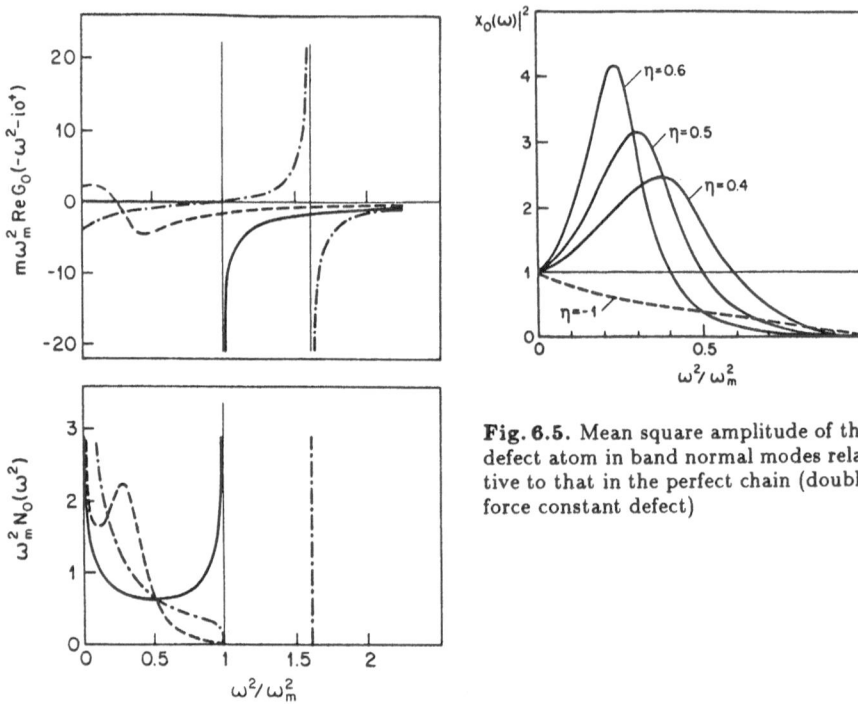

Fig. 6.5. Mean square amplitude of the defect atom in band normal modes relative to that in the perfect chain (double force constant defect)

Fig. 6.4. Behavior of $\mathrm{Re}\,\{G_0(-\omega^2 - i0^+)\}$ and of $N_0(\omega^2) = (m/\pi)\,\mathrm{Im}\,\{G_0(-\omega^2 - i0^+)\}$ for a monatomic linear chain with a double force constant defect. — $f_0 = f$ ($\eta = 0$, pure chain); - - - - $f_0 = f/2$ ($\eta = 1/2$); $\cdot$-$\cdot$-$\cdot$-$\cdot$ $f_0 = 2f$ ($\eta = -1$)

$$\frac{|\chi_0(\omega)|^2}{|\chi_0(\omega)|^2_{\eta=0}} = \frac{1 - \dfrac{\omega^2}{\omega_m^2}}{1 - \left(\dfrac{1+3\eta}{1-\eta}\right)\dfrac{\omega^2}{\omega_m^2} + \dfrac{4\eta}{(1-\eta)^2}\left(\dfrac{\omega^2}{\omega_m^2}\right)^2} \quad , \quad 0 < \omega < \omega_m \ .$$

$$(6.47)$$

As shown in Fig. 6.5, the resonance shifts to lower frequencies and its width becomes narrower as the defect force constant decreases ($\eta \to 1$). For $\eta < 0$, the mean square amplitude of the defect in a band mode is smaller than that of an atom in the perfect chain.

6.4 The Random Binary Mass Chain

6.4.1 General Features

In the last section we have shown that the effects of a single defect can be calculated exactly. Similar calculations can be carried out for any small number of defects in an infinite perfect crystal. As soon as the concentration of defects is non-zero, however, such calculations become intractable. To obtain a theoretical description of the vibrational properties of disordered systems, we therefore have to introduce approximations. Two important classes of such approximation methods will be introduced in the following subsections. To simplify the discussion, we shall restrict the corresponding calculations to a simple model system, the random binary-mass chain, although the approximation schemes themselves are rather general. Their extensions to treat more complicated types of disorder as well as higher dimensional systems are discussed in Sect. 6.5.

A random binary-mass chain is characterized by the values of the two masses, m_A and m_B, and by their respective concentrations, c and $1 - c$. Its vibrational density of states, $N(\omega^2)$, can be parametrized by the concentration c, by the relative mass defect ε,

$$\varepsilon \equiv \frac{m_B - m_A}{m_B} \; , \tag{6.48}$$

and by a normalizing frequency ω_m,

$$\omega_m^2 \equiv 4\frac{f}{m_B} \; . \tag{6.49}$$

Despite the difficulties to treat disordered systems, the vibrational spectra of random binary mass chains, and of other one- two- and three-dimensional model systems, are very accurately known. This is due to extensive numerical calculations on large finite disordered chains and lattices [6.13, 14]. An excellent review of these computer experiments has been given by *Dean* [6.15]. We therefore restrict our discussion to a few important aspects of these numerical calculations. The most striking feature in the numerically determined frequency spectra of binary mass-disordered chains is their extremely irregular (spiky) nature at high frequencies outside the heavy mass frequency band (Fig. 6.6). The spikes reflect the appearance of localized modes associated with isolated light masses or with particular clusters of light and heavy masses.

For $c = 0.2$, the monatomic host continuum of the heavy masses is still clearly distinguishable. Its erosion near the upper end continues as the concentration of light masses is increased, while more and more peaks

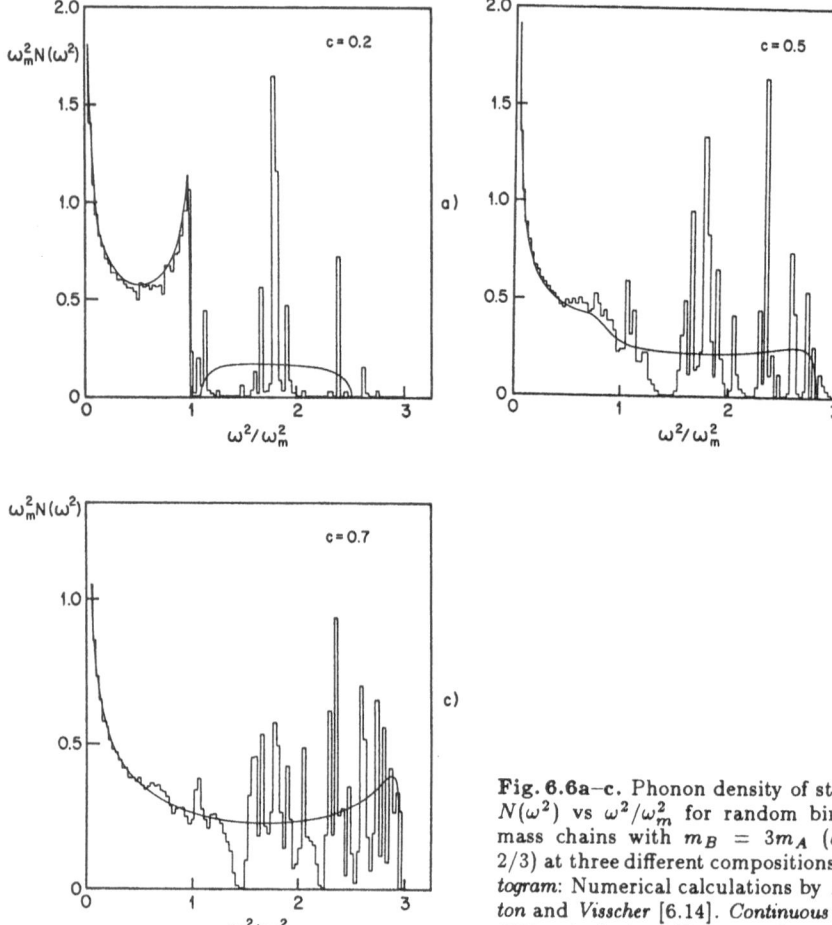

Fig. 6.6a–c. Phonon density of states $N(\omega^2)$ vs ω^2/ω_m^2 for random binary mass chains with $m_B = 3m_A$ ($\varepsilon = 2/3$) at three different compositions. *Histogram*: Numerical calculations by *Payton* and *Visscher* [6.14]. *Continuous line*: CPA calculations (Sect. 6.4.2)

appear at higher frequencies. As $c \rightarrow 1$, the spectrum then approaches the monatomic continuum of the light masses.

A similar spiky high-frequency structure of the spectrum is also observed for simple cubic lattices if the concentration of light masses is below the percolation critical concentration [6.8, 14]. Above the percolation threshold, however, most of the light atoms are connected together in an infinite cluster, so that the structures associated with isolated clusters become less important (see also Sect. 6.5).

The spiky nature of the spectrum is also much less pronounced if instead of a binary mass distribution we choose a broader or even continuous distribution of masses [6.15].

6.4.2 The Coherent Potential Approximation

One of the most important analytical approximations for disordered systems is the so-called coherent potential approximation (CPA) [6.5, 8, 9, 16], a simple, but very powerful, self-consistent effective medium approximation. In the following, we shall present a simple, intuitive derivation of the CPA for random binary mass chains. A more formal and general derivation of the CPA is sketched in Sect. 6.5.

We consider a harmonic linear chain with a random binary distribution of masses, but with constant nearest-neighbor force constants f. The disorder is thus characterized by a probability density of the form

$$\varrho(m) = c\delta(m - m_A) + (1 - c)\delta(m - m_B) \ , \tag{6.50}$$

i.e. the two masses m_A and m_B occur with probabilities c and $1 - c$, respectively.

We now wish to represent the random chain by a non-random, effective chain in which all masses have the same value, m_{eff}, which may be frequency dependent. In the CPA, this effective mass is determined in a self-consistent manner according to the following procedure.

We consider a particular mass m ($m = m_A$ or m_B) of the random chain, and replace all the other masses by m_{eff}. The calculation of the diagonal Green function $G_0(m, m_{\text{eff}}; z)$ corresponding to a stimulation of mass m then becomes a single defect problem, and according to Sect. 6.2.1 we obtain

$$G_0(m, m_{\text{eff}}; z) = \frac{1}{mz + 2X_{\text{eff}}} \ , \quad \text{where} \tag{6.51}$$

$$X_{\text{eff}} = \frac{1}{2}m_{\text{eff}}\left[z^{1/2}\left(4\frac{f}{m_{\text{eff}}} + z\right)^{1/2} - z\right] \ . \tag{6.52}$$

As a self-consistency requirement we now demand that the average of $G_0(m, m_{\text{eff}}; z)$ with respect to the mass distribution $\varrho(m)$ should be equal to the diagonal Green function $G_0^{\text{eff}}(z) \equiv G_0(m_{\text{eff}}, m_{\text{eff}}; z)$ of the effective chain, i.e.

$$cG_0(m_A, m_{\text{eff}}; z) + (1 - c)G_0(m_B, m_{\text{eff}}; z) = G_0(m_{\text{eff}}, m_{\text{eff}}; z) \ . \tag{6.53}$$

The self-consistency equation (6.53) can be written in the form

$$c\frac{m_A - m_{\text{eff}}}{1 + (m_A - m_{\text{eff}})zG_0^{\text{eff}}(z)} + (1 - c)\frac{m_B - m_{\text{eff}}}{1 + (m_B - m_{\text{eff}})zG_0^{\text{eff}}(z)} = 0 \ , \tag{6.54}$$

where

$$G_o^{\text{eff}}(z) = \cfrac{1}{m_{\text{eff}} z^{1/2} \left(4 \cfrac{f}{m_{\text{eff}}} + z \right)^{1/2}} \quad , \tag{6.55}$$

and leads to a cubic equation for the (complex and z-dependent) effective mass $m_{\text{eff}}(z)$.

Once $m_{\text{eff}}(z)$ is determined, the CPA result for the vibrational density of states of the random binary mass chain is given by the density of states of the corresponding effective chain, i.e.

$$N(\omega^2) = \frac{1}{\pi} \text{Im} \{ m_{\text{eff}}(z) G_0^{\text{eff}}(z) \}_{z = -\omega^2 - i0^+} \quad . \tag{6.56}$$

Some explicit results are presented in Fig. 6.6. The comparison with numerical calculations of *Payton* and *Visscher* [6.14] shows that the CPA gives a good overall description of the frequency- and concentration-dependence of $N(\omega^2)$, but also that it is not capable to reproduce the spike structure in the high-frequency region. The reason is that the spikes are due the response of isolated light masses or of particular small clusters, and that the simple CPA does not take into account such local environment effects.

We note, however, that the CPA leads to a surprisingly accurate description in situations where local environment effects are less important, e.g. in higher dimensional systems above the percolation threshold for the light masses (Sect. 6.5), or in disordered systems with a broad distribution of masses. It can further be shown that in the dilute limits ($c \to 0$ and $c \to 1$) the CPA gives a result which is exact to order c or $1 - c$, respectively.

6.4.3 Cluster Approximations

In the CPA, the effects of disorder are approximated by considering a single mass embedded in an effective medium whose parameters are then determined self-consistently. As a socalled single-site approximation, the CPA can therefore not reproduce effects of disordered systems that arise from the behavior of clusters of masses.

To take into account such local environment effects, various cluster extentions of the single-site CPA have been proposed [Ref. 6.9, Sect. 2.3.5]. However, many of these attempts suffer from certain limitations. They do, for example, not lead to properly analytic Green functions, fail to reproduce the correct dilute limits, are not self-consistent, or do not preserve the translational invariance of the reference crystal [6.9]. Nevertheless, such cluster-CPA's are capable to reproduce much of the spiky structure of the vibrational density of states of mass-disordered chains. As an example, the

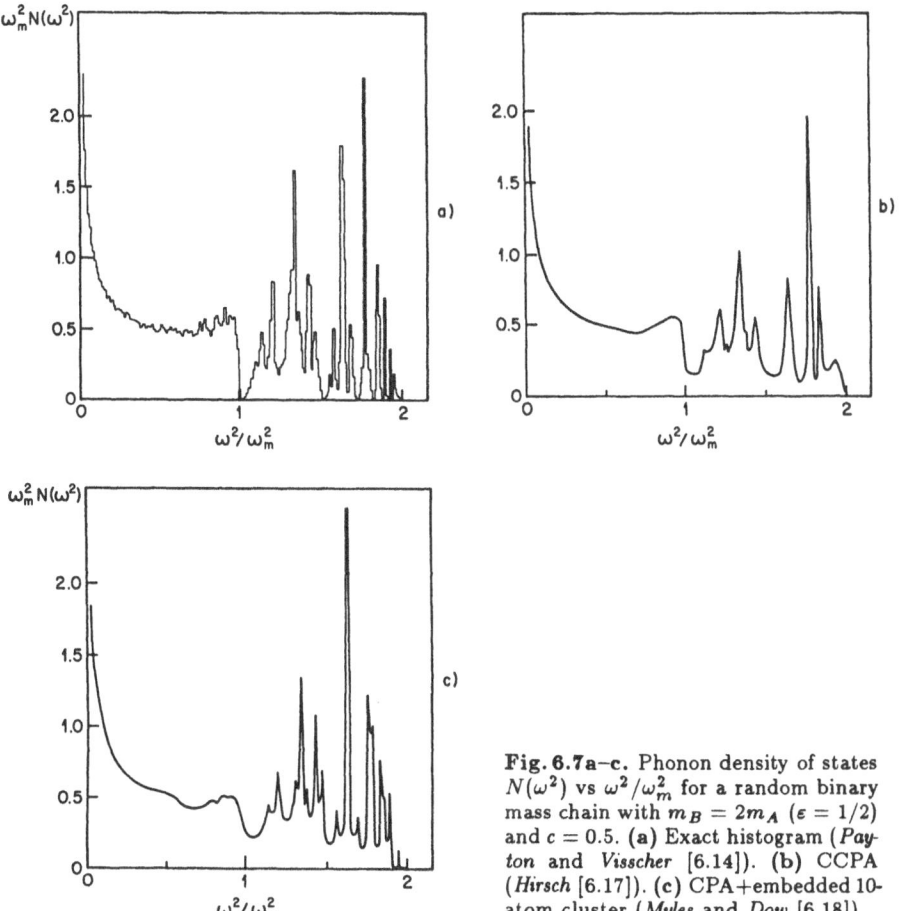

Fig. 6.7a–c. Phonon density of states $N(\omega^2)$ vs ω^2/ω_m^2 for a random binary mass chain with $m_B = 2m_A$ ($\varepsilon = 1/2$) and $c = 0.5$. (a) Exact histogram (*Payton* and *Visscher* [6.14]). (b) CCPA (*Hirsch* [6.17]). (c) CPA+embedded 10-atom cluster (*Myles* and *Dow* [6.18])

results of a 7-site cellular coherent potential approximation (CCPA) [6.17] are shown in Fig. 6.7b.

The evaluation of cluster-CPA theories is numerically rather cumbersome. This is due to the fact that not only a single site, but a whole cluster of the disordered system is embedded into an effective medium that now has to be characterized by several effective parameters. The averaging procedure and the self-consistency condition (represented by a set of coupled self-consistency equations) therefore become much more complicated than in the single-site CPA.

For this reason, a number of non-selfconsistent cluster approximations have been proposed. They require much less numerical work than selfconsistent calculations, but nevertheless lead to quite comparable results. As an example, we consider the embedded cluster method of *Myles* and *Dow* [6.18]. This method is based on the assumption that a small cluster of

155

atoms embedded in a suitably chosen effective medium should reproduce the main features of the vibrational spectrum. The effective medium is thereby chosen a priori (and not determined self-consistently), and the density of states of the disordered system is approximated by the density of states of the embedded cluster, averaged over all cluster-configurations. Very good results are obtained with a CPA-effective medium (Fig. 6.7c).

6.4.4 Renormalization Group Approaches

A completely different, and very promising method to determine the density of states in one-dimensional disordered systems has recently been introduced by *Goncalves da Silva* and *Koiller* [6.19]. The method is based on a real-space renormalization procedure applied to the equations of motion.

To describe this procedure, we first consider a uniform harmonic chain, i.e. $m_i = m$ and $f_{i,i+1} = f$ for all values of i. The equations of motion for the Green functions $G_i(z)$, (6.10), can then be written in the form

$$(mz + 2k^{(0)})G_i = f^{(0)}G_{i-1} + f^{(0)}G_{i+1} + \delta_{i0} \; , \tag{6.57}$$

where

$$k^{(0)} = f^{(0)} = f \; . \tag{6.58}$$

As renormalization procedure we choose a simple decimation of the one-dimensional lattice, i.e. we eliminate all odd-numbered sites. We then obtain

$$(mz + 2k^{(1)}G_i) = f^{(1)}G_{i-2} + f^{(1)}G_{i+2} + \delta_{i0} \; , \tag{6.59}$$

with

$$f^{(1)} = \frac{f^{(0)2}}{mz + 2k^{(0)}} \quad \text{and} \tag{6.60a}$$

$$2k^{(1)} = 2k^{(0)} - 2\frac{f^{(0)2}}{mz + 2k^{(0)}} \; . \tag{6.60b}$$

The renormalization procedure is thus seen to lead to a new chain, with renormalized "force constants" $f^{(1)}$ and $k^{(1)}$, respectively, and in which the sites are twice as far apart as in the original chain. As the new equations have the same structure as the original ones, the renormalization step may easily be repeated to eliminate more and more degrees of freedom. This is simply done by iterating (6.60a,b) for a fixed value of $z = -\omega^2 - i\varepsilon$, where $\varepsilon \to 0^+$ has to be chosen finite to ensure convergence.

After a few iterations the $f^{(n)}$ become small and very rapidly converge to zero, while $k^{(n)}$ tends to a limiting value, so that

$$G_0(-\omega^2 - i\varepsilon) = \lim_{n \to \infty} \frac{1}{m(-\omega^2 - i\varepsilon) + 2k)^{(n)}} \; . \tag{6.61}$$

By choosing ε small enough, the vibrational density of states of the infinite uniform chain,

$$N(\omega^2) = \frac{1}{\pi} \, \mathrm{Im} \, \{mG_0(-\omega^2 - i0^+)\} \ , \tag{6.62}$$

is therefore easily approximated to an arbitrary degree of accuracy.

Let us now try to apply the renormalization procedure to a binary mass-disordered chain. The force constants are again assumed to be uniform, $k^{(0)} = f^{(0)} = f$, but the masses m_i now randomly take the values m_A (with probability c) and m_B (with probability $1 - c$), respectively. After the first renormalization step, the renormalized force constants become

$$f_{i,i+1}^{(1)} = \frac{f^{(0)2}}{m_{i+1}z + 2k^{(0)}} \ , \tag{6.63a}$$

$$2k_i^{(1)} = 2k^{(0)} - \frac{f^{(0)2}}{m_{i-1}z + 2k^{(0)}} - \frac{f^{(0)2}}{m_{i+1}z + 2k^{(0)}} \ , \tag{6.63b}$$

i.e. they are now also random variables. To preserve the simple form of the renormalization procedure, *Goncalves da Silva* and *Koiller* [6.19] replaced (6.63a,b) by their averages with respect to the distribution of masses. The resulting approximate renormalization procedure leads to a density of states that reproduces the observed spiky structure, but the quantitative agreement with the "exact" computer-generated spectra is rather poor. The procedure can be improved to a certain extent by performing more renormalization steps before averaging, or by eliminating groups of sites rather than single sites [6.20]. *Robbins* and *Koiller* [6.21] also derived and discussed more elaborate renormalization schemes by combining the decimation procedure with conventional alloy averaging techniques.

It turns out, however, that the simple renormalization approach leads to much more accurate results if, instead of the force constants themselves, the *inverse* force constants are averaged after each renormalization step [6.22]. Intuitively, this seems to be a more natural averaging procedure, and we therefore propose the following renormalization scheme:

$$\frac{1}{f^{(1)}} = \left\langle \frac{1}{f_{i,i+1}^{(1)}} \right\rangle \ , \tag{6.64a}$$

$$\frac{1}{m_0 z + 2k^{(1)}} = \left\langle \frac{1}{m_0 z + 2k_i^{(1)}} \right\rangle \ , \tag{6.64b}$$

where $f_{i,i+1}^{(1)}$ and $k_i^{(1)}$ are defined in (6.63a and b), respectively, and where $\langle \ldots \rangle$ denotes the average with respect to the distribution of the masses

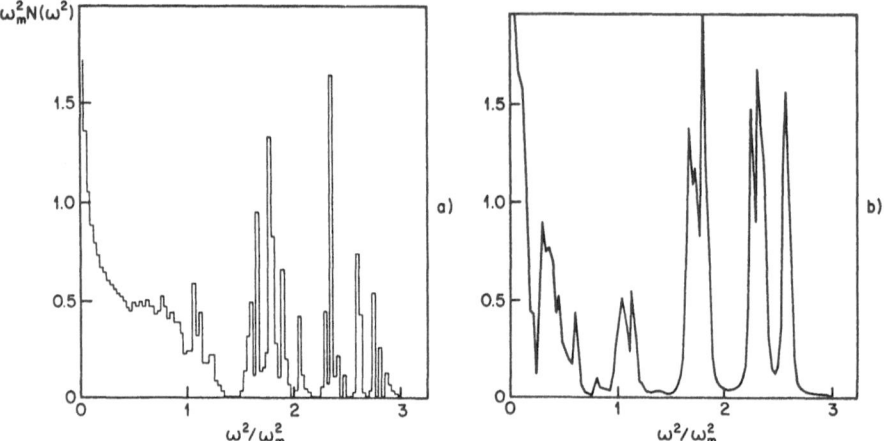

Fig. 6.8a,b. Phonon density of states $N(\omega^2)$ vs ω^2/ω_m^2 for a random binary mass chain with $m_B = 3m_A$ ($\varepsilon = 2/3$) and $c = 0.5$. (a) Exact histogram (*Payton* and *Visscher* [6.14]). (b) Renormalization method (6.64a,b and 65)

m_{i-1} and m_{i+1}, which independently take the values m_A and m_B (with probabilities c and $1 - c$, respectively). To calculate the density of states, finally, we have to add the contributions corresponding to the two possible values of the central mass m_0, and in analogy with (6.61 and 62) we then obtain

$$N(\omega^2) = \lim_{n \to \infty} \frac{1}{\pi} \mathrm{Im} \left\{ c \frac{m_A}{m_A z + 2k^{(n)}} \right.$$
$$\left. + (1 - c) \frac{m_B}{m_B z + 2k^{(n)}} \right\}_{z = -\omega^2 - i\varepsilon} \tag{6.65}$$

An example of such a calculation is shown in Fig. 6.8. The comparison with the "exact", computer-generated spectrum of *Payton* and *Visscher* shows that, apart from the low-frequency continuum, the major features of the spiky spectrum are reasonably well reproduced. Improvements of this simple renormalization approach, particularly in the low-frequency region, can again be obtained by performing several renormalization steps before averaging. Despite their limitations, the extreme numerical simplicity of such renormalization methods represents a big advantage, e.g. over cluster approximations.

6.5 Three-Dimensional Lattices and More Complicated Types of Disorder

In the previous sections, we have restricted our explicit calculations to simple one-dimensional chain systems. The approaches themselves, however, are quite general and can easily be generalized to treat two- or three-dimensional lattices and more complicated types of disorder.

To discuss some of these general aspects, it is convenient to use a more formal notation, and we start by writing the equations of motion for the Green functions in matrix form,

$$(Mz + \phi)G(z) = 1 \ , \tag{6.66}$$

where in this section M denotes the (diagonal) mass matrix, ϕ the force constant matrix, and 1 the unit matrix. The density of states, $N(\omega^2)$, is then given by

$$N(\omega^2) = \frac{1}{d\pi r N} \operatorname{Im} \{ \operatorname{Tr} \{ M^{1/2} G(-\omega^2 - i0^+) M^{1/2} \} \} \ , \tag{6.67}$$

where d is the dimensionality of the lattice, r the number of atoms in a unit cell, and $N(\to \infty)$ the number of unit cells in the lattice.

For a perfect crystal,

$$G^0(z) = (M^0 z + \phi^0)^{-1} \tag{6.68}$$

and $N^0(\omega^2)$ can be obtained by diagonalizing the (translationally invariant) force constant matrix ϕ^0 (or the dynamical matrix $D^0 = M^{0\,1/2} \phi^0 M^{0\,1/2}$, respectively). To calculate $G(z)$ for a system with defects, we introduce the perturbation matrix $V(z)$,

$$V(z) = (M^0 - M)z + (\phi^0 - \phi) \ , \tag{6.69}$$

and obtain the Dyson equation

$$G(z) = G^0(z) + G^0(z)V(z)G(z) \ . \tag{6.70}$$

If we introduce the T-matrix $T(z)$ by

$$T(z)G^0(z) = V(z)G(z) \ , \tag{6.71}$$

we have

$$G(z) = G^0(z) + G^0(z)T(z)G^0(z) \ , \tag{6.72}$$

with

$$T(z) = [1 - V(z)G^0(z)]^{-1}V(z) \ . \tag{6.73}$$

We now assume that $V(z)$ can be written as a sum of single-site contributions,

$$V(z) = \sum_i V^i(z) \ . \tag{6.74}$$

It follows that

$$T(z) = \sum_i T^i(z) \quad \text{with} \tag{6.75}$$

$$T^i(z) = t^i(z) + t^i(z)G^0(z) \sum_{j \neq i} T^j(z) , \tag{6.76}$$

where

$$t^i(z) = [1 - V^i(z)G^0(z)]^{-1}V^i(z) \tag{6.77}$$

describes the complete scattering by the ith defect.

In the case of a single defect, labelled by $i = i_0$, we have

$$V^i(z) = 0 \quad \text{if} \quad i \neq i_0 , \tag{6.78}$$

so that the exact expression for $G(z)$ simply becomes

$$G(z) = G^0(z) + G^0(z)t^{i_0}(z)G^0(z) \quad \text{with} \tag{6.79}$$

$$t^{i_0}(z) = [1 - V^{i_0}(z)G^0(z)]^{-1}V^{i_0}(z) . \tag{6.80}$$

Similarly, $G(z)$ can be evaluated exactly for any small number of defects. As soon as we have a finite concentration of defects, however, this is no longer possible and we have to use approximate methods.

In disordered systems, the quantity of interest is the configurationally averaged Green function, $\langle G(z) \rangle$. The simplest approximation is obtained by neglecting all correlations in the evaluation of

$$\langle G(z) \rangle = G^0(z) + G^0(z)\langle V(z)G(z) \rangle , \tag{6.81}$$

so that

$$\langle G(z) \rangle \cong G^0(z) + G^0(z)\langle V(z) \rangle\langle G(z) \rangle , \quad \text{or} \tag{6.82}$$

$$\langle G(z) \rangle \cong [1 - G^0(z)\langle V(z) \rangle]^{-1}G^0(z) . \tag{6.83}$$

This is the so-called virtual crystal approximation (VCA). A somewhat better approximation, the average t-matrix approximation (ATA), is obtained from (6.72 and 75–77) by decoupling the averages of products of $t^i(z)$-matrices. Both the VCA and the ATA are non-selfconsistent approximations [6.8,9], and we shall again concentrate our discussion on the self-consistent coherent potential approximation (CPA).

In the CPA we try to describe our disordered system by a non-random effective system, i.e. we require

$$\langle G(z) \rangle = G_{\text{eff}}(z) = [M_{\text{eff}}(z)z + \phi_{\text{eff}}(z)]^{-1} . \tag{6.84}$$

To determine $M_{\text{eff}}(z)$ and $\phi_{\text{eff}}(z)$ selfconsistently, we observe that in analogy with (6.72) we can write

$$G(z) = G_{\text{eff}}(z) + G_{\text{eff}}(z)T_{\text{eff}}(z)G_{\text{eff}}(z) \; , \tag{6.85}$$

where

$$T_{\text{eff}}(z) = [1 - V_{\text{eff}}(z)G_{\text{eff}}(z)]^{-1}V_{\text{eff}}(z) \; , \tag{6.86}$$

with

$$V_{\text{eff}}(z) = [M_{\text{eff}}(z) - M]z + [\phi_{\text{eff}}(z) - \phi] \; . \tag{6.87}$$

From (6.84 and 85) it follows that the effective system in principle is determined by

$$\langle T_{\text{eff}}(z)\rangle = 0 \; . \tag{6.88}$$

In this general form, however, the problem is as complicated as the original problem. With a simple (i.e., tractable) effective medium, (6.88) can only be satisfied approximately. We again assume that $V_{\text{eff}}(z)$ can be written as a sum of single-site contributions,

$$V_{\text{eff}}(z) = \sum_i V^i_{\text{eff}}(z) \; , \quad \text{so that} \tag{6.89}$$

$$T_{\text{eff}}(z) = \sum_i T^i_{\text{eff}}(z) \; , \tag{6.90}$$

$$T^i_{\text{eff}}(z) = t^i_{\text{eff}}(z) + t^i_{\text{eff}}(z)G_{\text{eff}}(z) \sum_{j\neq i} T^i_{\text{eff}}(z) \; , \tag{6.91}$$

$$t^i_{\text{eff}}(z) = [1 - V^i_{\text{eff}}(z)G_{\text{eff}}(z)]^{-1}V^i_{\text{eff}}(z) \; . \tag{6.92}$$

To approximate $\langle T^i_{\text{eff}}(z)\rangle$, we neglect all correlations between the scattering at different sites, i.e. we write

$$\langle T^i_{\text{eff}}(z)\rangle = \langle t^i_{\text{eff}}(z)\rangle + \langle t^i_{\text{eff}}(z)\rangle G_{\text{eff}}(z) \sum_{j\neq i} \langle T^j_{\text{eff}}(z)\rangle \; . \tag{6.93}$$

Equation (6.88) then leads to

$$\langle t^i_{\text{eff}}(z)\rangle = 0 \; , \quad \text{i.e.} \tag{6.94}$$

$$\langle [1 - V^i_{\text{eff}}(z)G_{\text{eff}}(z)]^{-1}V^i_{\text{eff}}(z)\rangle = 0 \; . \tag{6.95}$$

This is the general formal expression for the selfconsistency equation of the single-site CPA. It determines an effective system that approximates the behavior of the original disordered system.

In Fig. 6.9, the CPA results for the phonon spectra of binary mass-disordered simple cubic lattices are compared with the corresponding exact computer calculations. The CPA is seen to lead to an extremely accurate description in the three-dimensional case where the spectra are much less spiky than in one dimension. Pronounced spikes are only observed below the percolation concentration for the light masses, and there the CPA again fails to reproduce the detailed structure. The selconsistency equation (6.95) is valid for arbitrary site-diagonal disorder, i.e. for an arbitrary distribution of masses. However, the CPA is easily adapted to treat other types of disorder. In the case of force constant disorder, for example, we merely have to represent $V_{eff}(z)$ as a sum over single-bond contributions. The self-consistency equation then has exactly the same form as (6.95), with $V_{eff}^{i}(z)$ replaced by the single-bond potential $V_{eff}^{ij}(z)$.

Broad distributions of masses or force constants lead to rather smooth phonon spectra even in one dimension, as is demonstrated by the example in Fig. 6.10. In such situations, the CPA represents a very accurate approximation. From the neglect of correlations in (6.93), however, it is obvious that multiple-site scattering is not treated correctly by the single-site or single-bond CPA. Improved approximations can be obtained from the construction of cluster-CPA's which, by an appropriate decomposition of $V_{eff}(z)$, treat the scattering within certain clusters of sites correctly. Such extensions of the

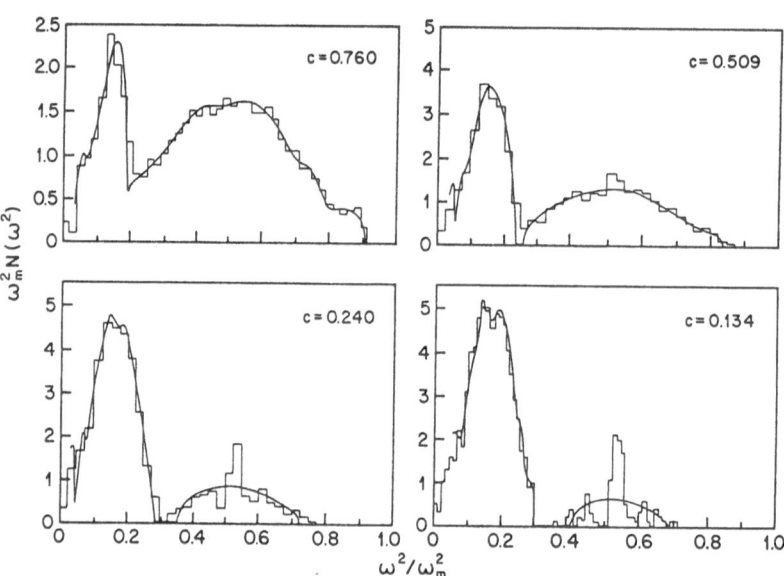

Fig. 6.9. Phonon density of states $N(\omega^2)$ vs ω^2/ω_m^2 for binary mass-disordered simple cubic lattices with $m_B = 3m_A$ ($\varepsilon = 2/3$) and compositions as indicated. The CPA-results are compared with computer generated histograms by *Payton* and *Visscher* [6.14]. After [6.16]

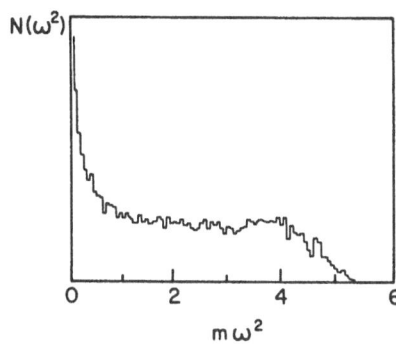

Fig. 6.10. Phonon density of states for a monatomic chain with force constants distributed uniformly between $f = 1/2$ and $f = 2/3$. After [6.15]

CPA are particularly important for the description of systems with mixed mass and force constant disorder, correlated disorder, or with further than nearest neighbor interactions.

An alternative class of approximations that include correllations between different sites has been discussed in Sect. 6.4.4. These approaches are based on a real-space renormalization procedure and require extremely little computational effort. In one dimension they yield the exact density of states in the ordered limits, and are easily adapted to treat very general types of disorder [6.21, 23]. Generalizations to higher dimensions seem possible, but have not yet been performed. We note, however, that in higher dimensions already the renormalization step can only be carried out approximately.

Up to now we have only considered systems with compositional or substitutional disorder, such as alloys and mixed crystals, or crystals with substitutional impurities. There exists, however, a second broad class of disordered systems: systems with *structural disorder*. Glassy and amorphous materials, for example, and superionic conductors belong to this class. We shall not attempt to describe the corresponding theoretical approaches here but merely mention that they include the investigation of large clusters by numerical methods, and the use of analytical methods to study the behavior of small clusters of atoms embedded in an effective medium [Ref. 6.9, Chap. 3]. Superionic conductors are discussed in detail in Chap. 7.

6.6 Experimental Results and Comparison with Theoretical Calculations

In the previous sections, we have often used the results of computer simulations to test the accuracy of theoretical approaches. In real materials, the effects of defects and disorder are most conveniently studied via one-phonon infrared absorption, Raman and inelastic neutron scattering experiments. To conclude this chapter on phonons in disordered systems, let us therefore

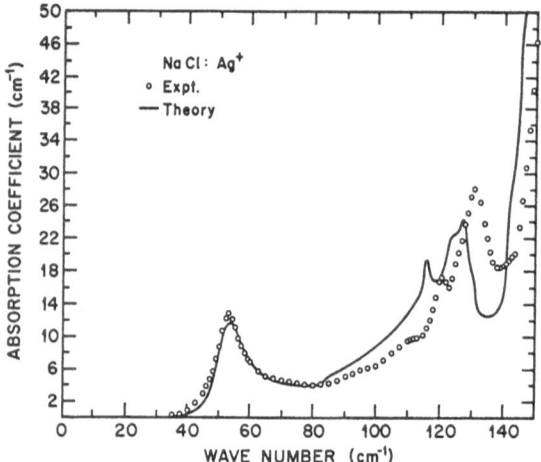

Fig. 6.11. Calculated infrared absorption due to Ag$^+$ in NaCl (—) compared with 7 K experimental data (o o o). After [6.24]

have a look at a few representative experiments where comparison has been made with correspoding theoretical calculations.

A careful study of the defect-induced far-infrared and first-order Raman spectra due to Ag$^+$ ions in NaCl has been made by *Montgomery* et al. [6.24]. Some of their beautiful results are reproduced in Figs. 6.11 and 12. The theoretical calculations are based on the assumption of effective force-constant changes which are then determined by a fit to the experimental data. The good quantitative agreement between the theoretical and experimental spectra could only be obtained if the force constant changes were assumed to depend on symmetry. This could be understood semi-quantitatively in terms of symmetry-dependent Coulomb contributions to the force-constant changes resulting from lattice relaxation. *Tsunoda* et al. [6.25] have studied the one-phonon coherent neutron scattering intensity in disordered Ni$_{1-c}$Pt$_c$ alloys. Their results for $c = 0.05$ are reproduced in Fig. 6.13. As a function of frequency, the scattering intensity exhibits a double peak structure which is due to resonant modes of the heavier Pt-atoms in the Ni host-crystal ($M_{Pt}/M_{Ni} = 3.22$). The CPA calculations only include single-site mass disorder effects and do not contain any fitting parameters. The agreement with the experimental data is quite reasonable, and the remaining discrepancies suggest the importance of force constant changes and of clustering effects. A detailed discussion of the data presented in this section, as well as of many more experimental studies performed on disordered materials, are presented in [6.7–9].

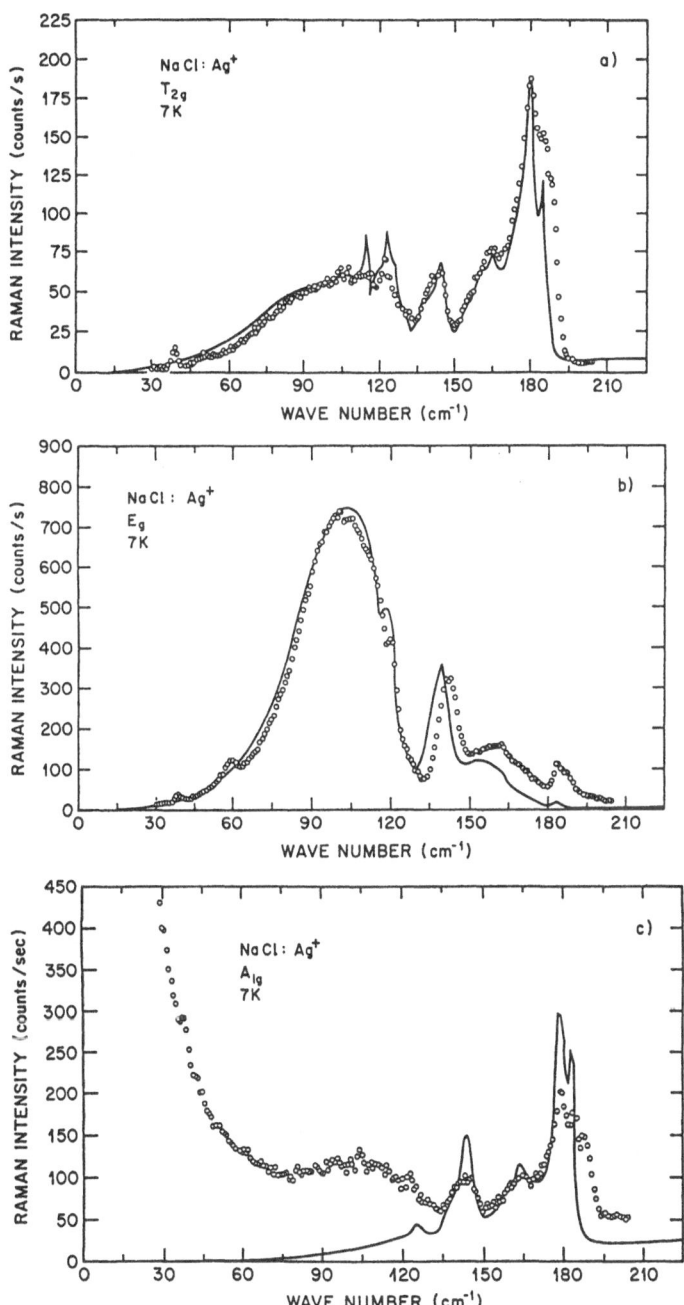

Fig. 6.12a–c. Calculated first-order Raman scattering due to Ag$^+$ in NaCl (—) compared with 7 K experimental data, for the T_{2g}, E_g, and A_{1g} modes, respectively. After [6.24]

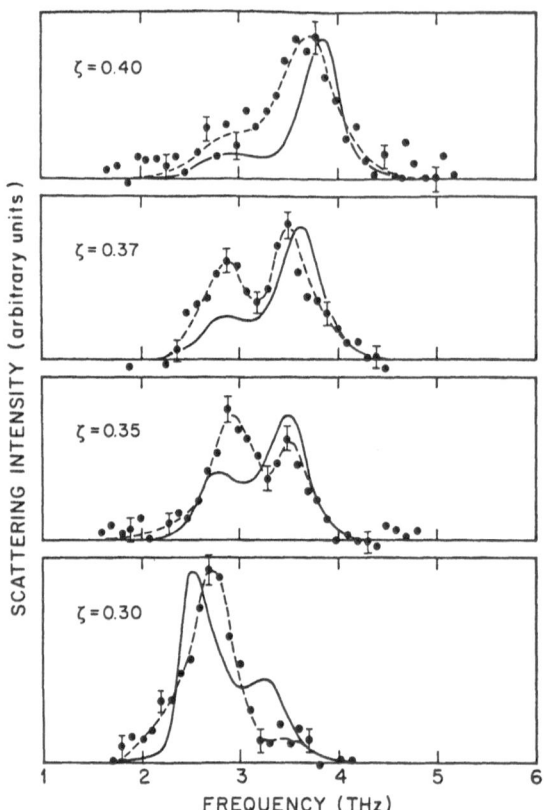

Fig. 6.13. Coherent neutron scattering intensity close to the resonance frequency for the [00ξ] transverse branch of $Ni_{0.95}Pt_{0.05}$ (ξ: reduced wave vector). (—) indicate the CPA line shapes and include the instrumental resolution function. After [6.25]

7. Ion Dynamics in Superionic Conductors

Superionic conductors (SIC) are solid electrolytes which have very high ionic conductivities comparable to those of liquid electrolytes. After a general introduction into the subject in Sects. 7.1, 2 the ionic interactions and dynamical models are discussed in Sect. 7.3. The simplest single-particle model which describes both the diffusive and oscillatory behaviour is the Brownian motion of a particle in a periodic potential treated in Sect. 7.4. Disorder-induced infrared absorption and Raman scattering is discussed in Sect. 7.5. Section 7.6 is devoted to a discussion of phonons in α-AgI and other SIC. Finally, in Sect. 7.7 a discussion is given to the lattice dynamical calculation of jump frequencies for ions in crystals with small defect concentrations as well as in SIC.

7.1 General Aspects of Superionic Conductors

The so-called superionic conductors (SIC) are characterized by an almost liquid-like mobility of one of the ionic species. For instance, in the high temperature $(T > 147°\,C)$ α-phase of AgI the mobility of the Ag^+ ions approaches the atomic mobility in molten metallic silver. As a consequence ionic conductivities of the order of $1\ (\Omega\,cm)^{-1}$ become possible. This is comparable with aqueous electrolytes, and hence the materials are often referred to as *solid electrolytes*. Solid electrolytes are of rapidly growing technical importance in energy storage and conversion schemes. From a theoretical point of view SIC are of great interest because these systems possess properties that are common to both solids and liquids. In fact, the arrangement of the mobile ions (Ag^+ in α-AgI) within the ordered framework (I^- in α-AgI) is frequently described as a molten sublattice.

Normal ionic conductors, such as NaCl, have ionic conductivities $\sigma_i < 10^{-10}\ (\Omega\,cm)^{-1}$, while for SIC σ_i is in the range between 10^{-3} and 2 $(\Omega\,cm)^{-1}$. On the other hand, the electronic conductivity of a good metal is of the order of $\sigma_e \cong 10^6\ (\Omega\,cm)^{-1}$. Thus the ratio of σ_e for a metal to σ_i of the best SIC is of the order of $\sigma_e/\sigma_i \cong m_i/m_e \cong 10^6$, where m_i is the mass of an ion and m_e is the electronic mass. Figure 7.1 shows the temperature dependence of σ_e of some normal ionic conductors, such as NaCl, NaI, LiCl,

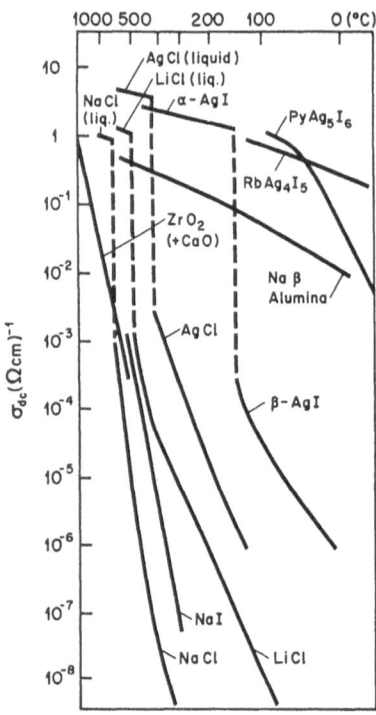

Fig. 7.1. Logarithm of the dc conductivities of some typical SIC plotted against $1/T$

AgCl, β-AgI, and of the SIC α-AgI, RbAg$_4$I$_5$, PyAg$_5$I$_6$, Na-β-Alumina and ZrO$_2$. For review articles about SIC the reader is referred to the literature [7.1–17].

7.2 Basic Facts and Examples of SIC

In order to obtain an insight into the conduction mechanism of the ions in SIC the knowledge of the crystal structure is of decisive importance. Consider the example of AgI. The normal ionic conductor β-AgI, stable below 147° C (Fig. 7.1) crystalizes in the well-ordered Wurtzite structure [Ref. 7.18, Fig. 3.7]. The basic reason for the abnormally high cation mobility of α-AgI (Fig. 7.1) was not recognized until 1934 when *Strock* [7.19] established that the silver ions in this phase are disordered and distributed over a large number of sites with a regular, crystalline arrangement of iodide ions. The structure of α-AgI is shown in Fig. 7.2. The iodide ions form a bcc lattice, while the two silver ions per unit cell dispose of 12 tetrahedral sites located at the faces of the unit cube. These sites are all equivalent and separated by a small potential barrier U, and hence the silver ions can jump from one site to neighbouring empty sites with a small activation energy $U \cong 0.1\,\mathrm{eV}$. This explains the high mobility μ of the conducting ions in SIC. In general, the ionic conductivity is given by

$$\sigma = zen\mu \quad , \tag{7.1}$$

where n is the concentration of mobile ions, ze their charge and μ their *mobility*. For Frenkel defects n is given by [7.20]

$$n = Bn_0\,e^{-E_f/2k_BT} \quad , \tag{7.2}$$

where E_f is the work required to remove an ion from a regular lattice site to an interstitial site (Fig. 7.3a), n_0 the concentration of normal lattice sites, and B an entropy factor. The mobility μ for ions in normal ionic conductors follows the *Arrhenius-relation*

$$\mu = C\frac{ze}{k_BT}a_0^2\nu_0\,e^{-U/k_BT} \quad . \tag{7.3a}$$

Here, ν_0 is an effective oscillation frequency, also called attempt frequency, which is typically comparable to a phonon frequency; ν_0 is a mean attack

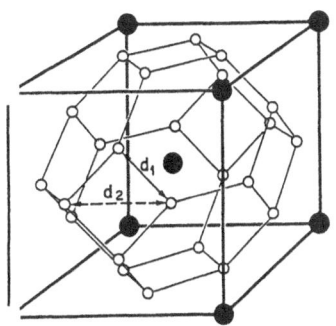

Fig. 7.2. Structure of α-AgI (stable above 147°C). The iodide ions ($\bullet\,\bullet\,\bullet$) form a bcc lattice; the two silver ions per unit cell dispose over 12 tetrahedral sites ($\circ\,\circ\,\circ$)

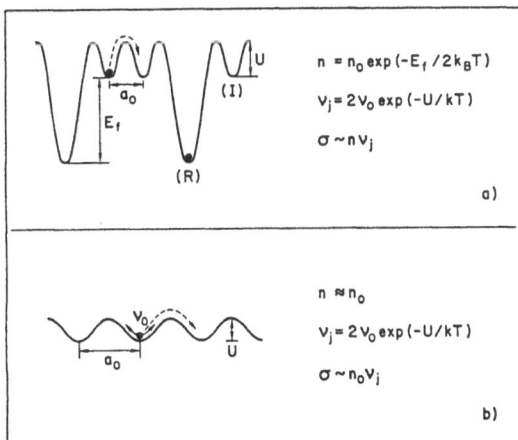

Fig. 7.3. (a) Potential energy of an ion in a normal ionic conductor; (R): regular lattice site, (I): interstitial site. (b) Potential energy of an ion in a SIC; ν_0 is the attempt frequency for jumps of the ions

frequency for ion jumps over the potential barrier U (Fig. 7.3). a_0 is the jump distance and C includes an entropy factor and a term depending on the crystal structure of the compound. The *jump frequency* is defined by

$$\nu_j = \nu_0 \, e^{-U/k_B T} \; , \tag{7.3b}$$

and the *diffusion constant* D is defined by the relation

$$\mu = ze D/k_B T \; . \tag{7.3c}$$

In normal ionic conductors E_f and U are large ($E_f \cong 2.0\,\mathrm{eV}$, $U \cong 0.85\,\mathrm{eV}$ in NaCl) and hence n and ν_j are small ($n \cong 10^{10}\,\mathrm{cm}^{-3}$, $\nu_j/\nu_0 \cong 10^{-10}$ at $150°$ C for NaCl). In medium ionic conductors E_f and U are already smaller ($E_f \cong 0.7\,\mathrm{eV}$, $U \cong 0.14\,\mathrm{eV}$ for β-AgI) and hence n and ν_j are considerably larger ($n \cong 10^{20}\,\mathrm{cm}^{-3}$, $\nu_j/\nu_0 \cong 2 \cdot 10^{-2}$ for β-AgI at $150°$C). In SIC there is no distinction between lattice sites and interstitial sites (Fig. 7.3b), hence $E_f \cong 0$ and $n \cong n_0 \cong 10^{22}\,\mathrm{cm}^{-3}$; in α-AgI the barrier energy $U \cong 0.14\,\mathrm{eV}$ as in β-AgI. In Fig. 7.3 the situation is schematically illustrated for the one-dimensional case. For SIC we have $\sigma = zen_0\mu$, that is, the conductivity is essentially limited by the mobility μ and not by the concentration of mobile ions. In β-AgI and α-AgI both, the jump frequencies ν_j and the mobilities μ are comparable since ν_0 and U are approximately equal; the drastic increase in σ at $147°$C is essentially due to the increase in n at the order-disorder transition temperature. Many of the known SIC may be classified under the following heading [7.11, 15, 17]:

   a) Silver- and copper-based compounds, e.g. AgI, $RbAg_4I_5$, Ag_3SI, $AgCrS_2$, CuI, CuBr, in which the disorder occurs in the silver and copper sublattices. Here the prototype material is AgI.
   b) Hexagonal compounds with the β-alumina structure $B_2O:nM_2O_3$ ($B =$ Na, K, Rb, Ag; $M =$ Al, Ga, Fe^{3+}; $n = 5$ to 11). The prototype here is sodium β-alumina, which, if available in stoichiometric form, would have the formula $Na_2O:11Al_2O_3$. Its superionic properties are determined by departures from stoichiometry.
   c) Fluorites, e.g. CaF_2, SrF_2, BaF_2. Much of the work on this group in recent years has centered on PbF_2.
   d) Defect stabilized ceramic oxides

$CaO:AO_2$ ($A =$ Zr, Hf, Th, Ce)

   e) Fluorides with the tysonite structure

AF_3 ($A =$ Y, Lu, Re) .

   f) Mixtures of a solid ionic conductor with fine particles of an insulating second phase can show a marked increase in conductivity compared to the pure, homogeneous system. An enhancement by 3 orders of magnitude in the Li conductivity of LiI after addition of Al_2O_3 particles has been observed, and similar observations have been reported for a variety of systems. The conductivity of such systems has recently been investigated from the point of view of percolation theory [7.104].

In order to obtain insight into the extremely complex properties of SIC, a wide range of experimental techniques is required. It is not sufficient to study only the static properties, such as the crystal structure and the dc conductivity $\sigma(0)$. *A deeper understanding of the physics of* SIC *requires knowledge of the full dynamics of ionic motion including the motion of the host lattice.* Such information can be obtained by studying the frequency dependent conductivity $\sigma(\omega)$ for frequencies ranging from the microwave to the infrared region, from Raman- and Brillouin scattering experiments as well as by inelastic neutron scattering techniques.

In this chapter we present a qualitative discussion of some illustrative experimental and theoretical studies of the dynamics of ions in SIC. There are two particular aspects of the ion dynamics in such systems which are of special interest. One is that of diffusion which will be dominating the response of the system when the frequency of the external perturbation is smaller than the jump frequency ν_j of the ions. For the diffusive motion of the ions, anharmonicity is of prime importance; in fact, diffusion can be regarded as the ultimate anharmonic event. In contrast to this the characteristic feature of the high frequency response is that of oscillations of ions with frequencies of the order of the attempt frequency ν_0 in a strongly disordered system. The latter frequency region may be considered to be dominated by phonons, in contrast to the diffusion-dominated region for $\nu \ll \nu_j$. It should be emphasized, however, that due to the strong disorder present in SIC, the basic concepts of solid-state physics – phonons, energy bands, Brillouin zones – must be, to some extent, modified or abandoned. If we continue to speak about phonons in such systems we are using this word in a loose sense, that is, we are referring to the anharmonic vibrations of ions in a structually disordered lattice for which the q-selection rules break down. Due to disorder, anharmonicity and possible coupling of oscillations (phonons) with the hopping motion of the ions, the "phonon" frequency spectra of SIC will be changed considerably as compared with that of the ordered or low-conductivity phase. Experimentally, these changes manifest themselves in disorder-induced infrared absorption and light scattering, for example, which gives rise to additional structures in the far-infrared conductivity $\sigma(\omega)$ and in the Raman spectra.

From the above considerations it follows that the motion of a single mobile ion can be pictured as that of a *diffusing oscillator* as shown schematically in Fig. 7.4. The mobile ion vibrates for a certain time τ_{res}, the *residence time* or *dwell time* in the *cage* formed by the nearest ions of the host lattice.

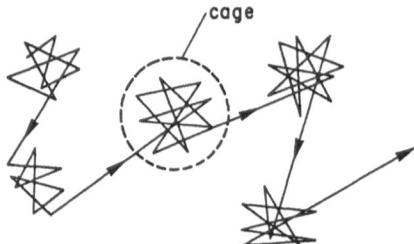

Fig. 7.4. The motion of a mobile ion in a SIC can be pictured as that of a diffusing oscillator; during the residence time τ_{res} the ion oscillates in its cage formed by the nearest neighbor ions of the host lattice and then jumps to a neighboring cage

It then leaves this cage and jumps into a neighbouring cage where it continuous to vibrate. The mean time an ion needs to fly from one cage to the next cage is called the *transit time* τ_{trans} or the flight time. The residence time is of the order of several vibrational periods ν_0^{-1}. For most SIC $\tau_{trans} \ll \tau_{res}$, but for the best SIC such as α-AgI or RbAg$_4$I$_5$, τ_{trans}/τ_{res} is estimated to be about 1/10 to 1/3 [7.21].

Due to the high temperatures and large masses involved in SIC the motion of the ions can be treated classically.

7.3 Ionic Interactions and Dynamical Models

The models which are used to describe SIC differ in whether certain interactions are neglected or considered and how these interactions are treated. In the following we introduce general interactions between the ions and give a qualitative discussion of certain models which are used to describe the ion dynamics of SIC. This discussion follows partly that given by *Geisel* [7.21].

In Fig. 7.5 the interactions are illustrated schematically for a given instantaneous configuration of the disordered crystal. The mobile ions are

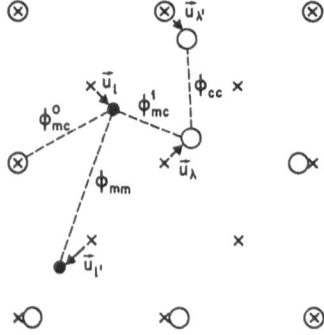

Fig. 7.5. Illustration of the interactions in (7.4). The mobile ions m (•••) are allowed to take any position, the cage ions c (ooo) vibrate around their equilibrium positions

denoted by m and the cage ions of the host lattice by c. The index l is used to number the mobile ions, while the index λ numbers the cage ions; R_l specifies the general position of the mobile ion l, and $R_\lambda = r_\lambda + u_\lambda$ gives the position of the cage ion λ, u_λ being its displacement with respect to its equilibrium position r_λ. The potential energy Φ contains interactions among all particles and we write

$$\begin{aligned}
\Phi = \Phi_{mc}^0(\ldots R_l, r_\lambda \ldots) + \Phi_{mc}^1(\ldots R_l, u_\lambda \ldots) \\
+ \Phi_{mm}(\ldots R_l, R_{l'} \ldots) + \Phi_{cc}(\ldots u_\lambda, u_{\lambda'} \ldots) \ .
\end{aligned} \tag{7.4}$$

For reasons discussed below we have split the interactions between mobile ions and cage ions into two terms, $\Phi_{mc} = \Phi_{mc}^0 + \Phi_{mc}^1$; Φ_{mm} and Φ_{cc} are the interactions between mobile ions and cage ions, respectively. Φ_{mc}^0 contains the interactions between the mobile ions with the cage ions located at their equilibrium positions, while Φ_{mc}^1 describes the interactions of the mobile ions with the cage ions at displaced positions. Thus Φ_{mc}^0 depends only on the continuous variables R_l, whereas Φ_{mc}^1 depends on both, R_l and R_λ.

It is certainly possible to expand Φ_{cc} and Φ_{mc}^1 in terms of the displacements u_λ of the cage ions. On the other hand, a similar expansion of Φ_{mc}^0, Φ_{mc}^1 and Φ_{mm} in terms of displacements u_l of the mobile ions is, in general, not possible, since a given mobile ion can not be associated with a fixed cage for $t > \tau_{res}$ (diffusive regime). However, if we are interested in the high-frequency "phonon-dominated" response only ($t < \tau_{res}$), and if we disregard the hopping motion, we can expand all interactions in (7.4) not only with respect to u_λ but also with respect to u_l (Fig. 7.5). Since the displacements of the mobile ions may be quite large ($u_l \lesssim r$, where r is the "radius" of the cage), it will, in general, be necessary to introduce anharmonic terms as well in the expansion of Φ_{mc}^0, Φ_{mc}^1 and Φ_{mm}. In this approximation the high-frequency motion of the ions in SIC can be reduced to the study of "phonons" in a structurally disordered crystal. This treatment is a good approximation if $\tau_{trans} \ll \tau_{res}$.

7.3.1 Free-Ion Model

One of the first attempts to treat the relatively free motion of the mobile ions in the best SIC was made by *Rice* and *Roth* [7.22]. The motion in the cage potential Φ_{mc}^0 is replaced by bound and free ion-like states which are separated by an energy gap. The other interactions in (7.4) are not considered explicitly. For a certain lifetime the ions are thermally excited and propagate with the velocity of free ions. The model leads to a Drude-behaviour for the frequency dependent conductivity $\sigma(\omega)$ (Fig. 7.8 for $k_B T/U \gtrsim 1$).

7.3.2 Hopping and Lattice Gas Models

In the lattice gas model it is assumed that the volume Ω of the crystal can be devided up into cells Ω_c ($c = 1 \ldots M$), such as to a good approximation a cell (cage) is either empty or occupied by one mobile ion only [7.23, 24]. Clearly, for a SIC there must be more cells than particles, $M > N$. In the lattice gas model the continuous motion of the mobile ions is replaced by a discontinuous motion from cell to cell. This is equivalent to the assumption of instantaneous jumps between the sites which is justified if $\tau_{trans} \ll \tau_{res}$. The potential Φ^0_{mc} and the interactions Φ^1_{mc} and Φ_{mm} are taken into account at the cell sites only or are meant as an average over the cell region. The potential barriers controlling diffusion are no longer incorporated in Φ^0_{mc}. In order to get ionic transport a transfer Hamiltonian is introduced, which produces jumps from one cell to another. The question *how* transport occurs can not be answered and part of the dynamics is anticipated by the transfer Hamiltonian. The problem is discussed in more detail by *Beyeler* et al. [7.23].

When the coupling Φ^1_{mc} between mobile ions and cage ion displacements $\boldsymbol{u}_\lambda$ is considered, it is often treated in terms of the small polaron concept [7.25-31]. Physically this allows for an accomodation of the cage to the site occupancy by mobile ions. In other words, the mobile ion carries its displacement field with it.

An advantage of the lattice gas model is that part of the interaction of Φ_{mm} among the mobile particles (that is the on-site interaction considered in the lattice gas model) can be treated more easily than in the continuous models discussed below.

7.3.3 Continuous Stochastic Models

Such models are usually introduced on a phenomenological basis. Here we quote only the underlying ideas. The mobile ions are assumed to move *continuously* in the cage potential Φ^0_{mc}. This potential is periodic with the periodicity of the host lattice and provides forces which depend on the positions of the mobile ions only (Fig. 7.6). The potential is strongly anharmonic. The restoring forces in the valley of the potential lead to oscillatory types of motion, the barriers control the diffusive motion. The vibrations of the cage described by Φ_{cc} are treated as a bath. Through the coupling Φ^1_{mc} they provide random forces and friction for the motion of the mobile ions. This is, of course, a simplification, in that the dynamical coupling between mobile ions and cage degrees of freedom has been taken into account only roughly. Interactions among the mobile ions described by Φ_{mm} are not treated explicitly, but are also considered to be incorporated in the bath.

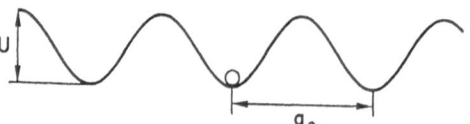

Fig. 7.6. Brownian motion of a particle in a periodic potential

The aim of these models is to treat the oscillatory and diffusive motions as illustrated in Fig. 7.4 simultaneously. These models are especially suited to describe ionic motion in the best SIC for which $\tau_{\text{trans}}/\tau_{\text{res}}$ is relatively large [7.32–38].

7.3.4 Continuum Models for the Hydrodynamic Region

In the long-wavelength and low-frequency region the mobile ions are described collectively by their density $n(\mathbf{r}, t)$ and current $\mathbf{j}(\mathbf{r}, t)$. They are coupled with the strain field of the crystalline cage. The theories [7.39–45] aim at a calculation of coupled liquid-cage fluctuations which are responsible for Brillouin scattering and inelastic neutron scattering at small momentum transfer (quasi-elastic scattering). A rigorous microscopic derivation based on the interaction energy (7.4) has been given by *Zeyher* [7.43, 44]. Of course, the continuum theories are valid for long-wavelength fluctuations which do not involve individual motion of particular ions, such that the individual treatment can be abandoned in favour of a collective treatment. On the basis of a continuum model, in which the crystalline cage is immersed in a viscous liquid, *Hayashi* et al. [7.41] and *Mizoguchi* [7.45] have studied the frequency-dependent conductivity and applied the results to α-AgI.

7.3.5 Molecular Dynamics Calculations

The inherent complexity of SIC makes it extremely difficult to treat their properties by analytical methods. Because of this, computer simulation methods have a major part to play in the theory of these materials. These calculations consider a limited number of anions and cations interacting with each other. If the many-body system is started in any reasonable initial condition and allowed to evolve freely according to Newton's equations of motion, it will attain the state characteristic of thermal equilibrium. In this state, it should resemble rather closely, both in its structural and dynamical properties, the real material studied experimentally, provided only that the forces between the ions are modelled realistically. It is possible to take into account all the ionic interactions contained in the potential energy (7.4), usually in the form of two-particle interactions. The form commonly assumed for the interaction energy between two ions consists of four terms, the Coulomb energy, the repulsion due to overlap of electronic charge clouds, the London dispersion energy, and the polarization self-energy of the ions which involves the ionic polarizabilities. Molecular dynamics calculations have been performed for CaF_2 by *Jacucci* and *Rahman*, and by *Dixon* and *Gillan* [7.46, 96]; for a computer simulation of ionic disorder in high-temperature PbF_2 the reader is referred to the study of *Walker* et al. [7.97]. α-AgI has been studied by *Schommers* [7.47], *Vashishta* and *Rahman* [7.48],

Fukumoto and *Ueda* [7.49], and by *Parinello* et al. [7.98]. The latter study is of particular interest since the $\alpha \rightarrow \beta$ phase transition is correctly reproduced by using a new molecular-dynamics technique which allows for a dynamical variation of both the volume *and* the shape of the unit cell. A review of the method applied to SIC has been given by *Gillan* [7.50].

7.4 Brownian Motion of a Particle in a Periodic Potential

The simplest single-particle model which describes both the diffusive and oscillatory behaviour is the Brownian motion of a particle in a periodic potential [7.32–37]. The corresponding Langevin equation is

$$m\ddot{x} + m\gamma\dot{x} - K(x) = f(t) \ , \tag{7.5}$$

where m is the mass of the particle, γ the damping, $K(x) = -\partial\Phi^0_{mc}/\partial x$ the periodic force orginating from the cage ions located at their equilibrium positions, see (7.4), and $f(t)$ a stochastic driving force. The damping γ and the random force $f(t)$ are due to interactions of the mobile ion with the vibrations of the cage ions and with other mobile ions, that is, they describe in a rough way the effects of the interactions Φ^1_{mc}, Φ_{cc} and Φ_{mm} in (7.4). For simplicity it is usually assumed that $f(t)$ has a white power spectrum [7.51], i.e.

$$\langle f(t)f(0)\rangle_T = 2m\gamma k_B T\delta(t) \ , \tag{7.6}$$

where $\langle \ldots \rangle_T$ denotes a thermal average, k_B the Boltzmann constant and T the temperature.

We assume that the cage potential $\Phi^0_{mc}(x)$ due to the rigid host lattice is

$$\Phi^0_{mc}(x) = -\frac{U}{2} \ \cos \frac{2\pi}{a_0} x \ , \tag{7.7}$$

where U is the potential barrier and a_0 the lattice constant (Fig. 7.6). Using (7.7), the Langevin equation (7.6) can be written in the form

$$m\ddot{x} + m\gamma\dot{x} + m\omega_0^2 \frac{a_0}{2\pi} \ \sin \frac{2\pi}{a_0} x = f(t) \ . \tag{7.8}$$

Here ω_0 is the oscillation frequency of the particle around the minima of $\Phi^0_{mc}(x)$ for small displacements x and is given by $\omega_0^2 = (2\pi)^2 U/2ma_0^2$. Two limiting cases are obvious. If the temperature is small compared with U, $k_B T \ll U$, the dynamic response of the particle becomes that of a damped harmonic oscillator, whereas in the opposite limit, $k_B T \gg U$ it becomes that of a freely diffusing particle. It is the aim of this section to give a qualitative discussion of the dynamic response for arbitrary temperatures and to compare the results with the observed frequency dependent conductivity $\sigma(\omega, T)$. For a derivation of the results the reader is referred to the literature [7.32, 35, 52].

It is possible to derive from (7.8) the velocity-velocity correlation function $\langle \dot{x}(t)\dot{x}(0)\rangle_T$. From this correlation function the mobility μ of the particle as a function of frequency and temperature can be evaluated via

$$\mu(\omega, T) = \frac{ze}{k_{\mathrm{B}}T} \int\limits_0^\infty dt\, e^{i\omega t} \langle \dot{x}(t)\dot{x}(0)\rangle_T \ , \tag{7.9}$$

where ze is the charge of the particle. Using a continued-fraction expansion technique one obtains [7.32, 53]

$$\mu(\omega, T) = \frac{ze}{m} \cfrac{1}{-i\omega + \gamma_0 + \cfrac{\Delta_1}{-i\omega + \gamma_1 + \cfrac{\Delta_2}{-i\omega + \gamma_2 +}}} \ . \tag{7.10}$$

For the case of the cosine potential (7.7) the terms γ_k and Δ_k up to $k = 2$ are $\gamma_0 = \gamma$, $\gamma_1 = 0$, $\gamma_2 = \gamma$, $\Delta_1 = \omega_0^2 \alpha$, $\Delta_2 = \omega_0^2 \beta$, with $\alpha = I_1(V)/I_0(V)$ and $\beta = \alpha^{-1} - V^{-1} - \alpha$; $I_p(V)$ is the modified Bessel function of order p [7.54] and $V = U/2k_{\mathrm{B}}T$. The ionic conductivity is then given by

$$\sigma(\omega, T) = ze\mu(\omega, T) \ . \tag{7.11}$$

Some exact results can be derived for the dc conductivity $\sigma_{\mathrm{dc}} = \sigma(\omega = 0, T)$ [7.32, 55–57].

(a) In the low temperature limit $(k_{\mathrm{B}}T \ll U)$ one obtains

$$\sigma(O, T) = (ze)^2 \frac{a_0^2}{2\pi k_{\mathrm{B}}T} \left[\left(\omega_0^2 + \frac{\gamma^2}{4} \right)^{1/2} - \frac{\gamma}{2} \right] e^{-2V} \ . \tag{7.12}$$

(b) In the limit of vanishing damping $(\gamma \to 0)$ one finds

$$\sigma(0, T) = (ze)^2 \frac{a_0}{(2\pi k_{\mathrm{B}}T)^{1/2}} \frac{e^{-2V}}{I_0(V)} \ . \tag{7.13}$$

(c) In the limit of large damping the result is

$$\sigma(0, T) = (ze)^2 \frac{1}{m\gamma} \frac{1}{I_0^2(V)} \ . \tag{7.14}$$

It is interesting to note that Arrhenius behaviour only holds in the limits of low temperature and small damping. Combining the limits (a) and (b) we obtain in fact

$$\sigma_{\mathrm{Arrh}} = (ze)^2 \frac{a_0 \nu_0}{k_B T} e^{-2V} \;, \tag{7.15}$$

where $\nu_0 = \omega_0/2\pi$ is the attempt frequency introduced in Sect. 7.1. By interpolating between the three different cases (a), (b), (c) one can obtain good approximations for $\sigma_{\mathrm{dc}}(T)$ for any values of V and γ. The damping is assumed to be proportional to T, $\gamma = \tilde{\gamma} k_B T$, because we are in the classical regime; the damping parameter is then $\tilde{\gamma}$. Figure 7.7 shows the qualitative behaviour of $\sigma_{\mathrm{dc}}(T)$. At low temperatures the ionic conductivity is activated and therefore increases with temperature. This trend stops at $V \cong 1$ (Fig. 7.7), where instead σ_{dc} begins to decrease up to the high temperature limit where the conductivity is Drude like. In this region the mobile ions do not feel anymore the potential, and the conductivity decreases due to the large damping in (7.14). The region in between the two dashed lines in Fig. 7.7 is the one appropriate for SIC.

We now turn to a discussion of $\sigma(\omega, T)$. For high temperatures $(k_B T \gg U)$, Δ_1 in (7.10) is negligible and one obtains

$$\sigma(\omega, T) = \frac{(ze)^2}{m} \frac{1}{-i\omega + \gamma} \;, \tag{7.16}$$

which describes the Drude behaviour of free ions. At low temperatures $(k_B T \ll U)$, Δ_2 in (7.10) is negligible with the result

$$\sigma(\omega, T) = \frac{(ze)^2}{m} \frac{1}{-i\omega + \gamma + \bar{\omega}_0^2/(-i\omega)} \;, \tag{7.17}$$

where $\bar{\omega}_0^2 = \alpha \omega_0^2$. This is the conductivity of a damped harmonic oscillator with an effective frequency $\bar{\omega}_0(T)$.

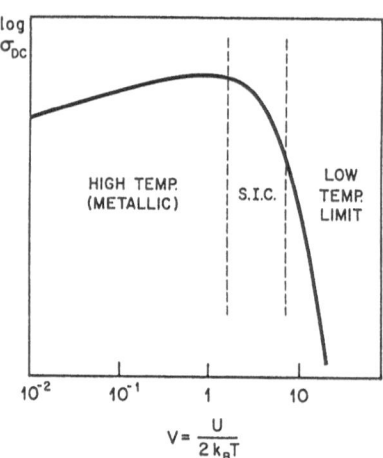

Fig. 7.7. Typical behaviour of σ_{dc} (arbitrary units) for a Brownian particle in a sinusoidal potential as a function of $V = U/2k_B T$ (U: barrier height) [7.37]

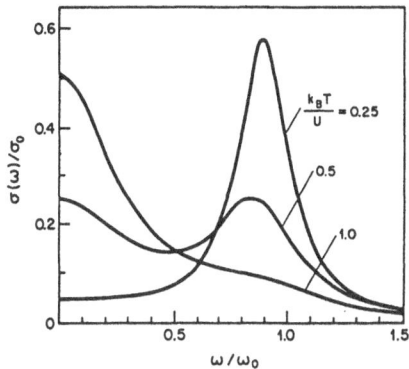

Fig. 7.8. Frequency dependent conductivity of a Brownian particle in a cosine potential at three different temperatures and small damping; U is the potential barrier in Eq. (7.7)

Figure 7.8 shows the real part of $\sigma(\omega, T)$ as obtained from (7.10, 11) for the cosine-potential (7.7) at three different temperatures [7.35]. In this figure $\sigma_0 = (ze)^2/m\gamma$ and $\gamma = \omega_0/2\pi$ (small damping). The transition from the oscillatory to the diffusive regime is clearly seen, as the temperature increases. At $k_B T/U = 0.25$, a strong oscillatory peak appears approximately at the frequency ω_0. At $k_B T/U = 1$ the spectrum is dominated by a central peak at $\omega = 0$, reflecting the diffusive motion of the ion over the potential barrier. At an intermediate temperature $k_B T/U = 0.5$, a pronounced minimum shows up in $\sigma(\omega)$ near $\omega \cong 0.5\,\omega_0$. For higher damping constants $(\gamma \gtrsim \omega_0/\pi)$ the transition between both regimes takes place without the appearance of a minimum between $\omega = 0$ and $\omega = \omega_0$.

An approximate expression for $\sigma(\omega, T)$ is obtained by truncating the continued fraction (7.10) after the second order and write for the conductivity

$$\sigma(\omega, T) \cong \frac{(ze)^2}{m} \cdot \frac{1}{-i\omega + \gamma_0 + \dfrac{\Delta_1}{-i\omega + R}} \ . \tag{7.18}$$

In this low order the remainder R may not be neglected. It is often approximated by a value independent of ω which is determined from the dc conductivity $\sigma(0, T)$. Equation (7.18) reproduces an oscillatory peak at $\omega = \omega_0$ at low temperatures and a diffusive peak at $\omega = 0$ at high temperatures much as in Fig. 7.8. However, it can not reproduce the simultaneous occurrence of both peaks at $k_B T/U = 0.5$ in Fig. 7.8. Apart from this case which shows up at low damping constant only, $Fulde$ et al. [7.32] have found that the truncated form (7.18) does not deviate significantly from higher-order continued fractions. We may therefore consider (7.18) a reasonable approximation for not too small values of the damping γ.

The relation (7.18) for $\sigma(\omega, T)$ has been established earlier in a similar form on a phenomenological basis [7.33]. The reasoning underlying the phenomenological approach is the following: If we look at the ion in a short time

179

interval $(t<\tau_{\text{res}})$, the particle motion obviously is described by a damped harmonic oscillator driven by a random force $f(t)$,

$$m\ddot{x} + m\gamma\dot{x} + m\bar{\omega}_0^2 x = f(t) \ . \tag{7.19}$$

At long times $(t>\tau_{\text{res}})$, however, the system is described by a diffusion equation:

$$m\ddot{x} + m\gamma'\dot{x} = f(t) \ . \tag{7.20}$$

We now introduce a generalized Langevin equation which holds for arbitrary times (or frequencies) and which is based on the memory function formalism [7.51, 58]:

$$m\ddot{x} + m\gamma\dot{x} + m\bar{\omega}_0^2 \int_0^t M(t-t')\dot{x}(t')dt' = f(t) \ . \tag{7.21}$$

The memory function $M(t)$ is formally a generalized damping [7.51, 58]. Oscillations and (7.19) are reproduced by the choice $M(t-t') = 1$, whereas diffusion and (7.20) are reproduced by the choice $M(t-t') = 0$. For intermediate times there should be a continuous transition from oscillatory to diffusive behaviour, i.e. $M(t-t')$ should decay continuously from 1 to 0 with increasing time. The simplest choice which satisfies this condition is

$$M(t-t') = e^{-(t-t')/\tau_{\text{C}}} \ , \tag{7.22}$$

where τ_c is the critical time between oscillatory and diffusive behaviour. This method provides an expression for $\sigma(\omega, T)$ equal to the truncated form of the continued-fraction expansion, (7.18), and one finds

$$\sigma(\omega, T) = \frac{(ze)^2}{m} \frac{1}{-i\omega + \gamma + \dfrac{\bar{\omega}_0^2}{-i\omega + 1/\tau_{\text{C}}}} \ . \tag{7.23}$$

The memory-function approach outlined above has been applied by *Brüesch, Pietronero, Straessler* and *Zeller* in several papers [7.30, 33, 34, 37]. These authors compared (7.23) with the conductivity $\sigma(\omega)$ of α-AgI derived from far-infrared reflectivity measurements. A fit of the data is shown in Fig. 7.9. The fit is good for $\omega \gtrsim 20\,\text{cm}^{-1}$. However, it is not possible to account for the pronounced shoulder observed in the region between ~ 7 and $\sim 20\,\text{cm}^{-1}$. In the model considered above the mobile ion moves in a rigid potential given by Φ_{mc}^0 in (7.4). Therefore, it completely neglects the polarizability associated with the displacements of the cage ions and the corresponding polaron effects. It can be shown that lattice polarizability can lead to structure in $\sigma(\omega)$ at the transition from diffusion-controlled to oscillation-controlled behaviour [7.29, 30]. Qualitatively this is to be expected since in the os-

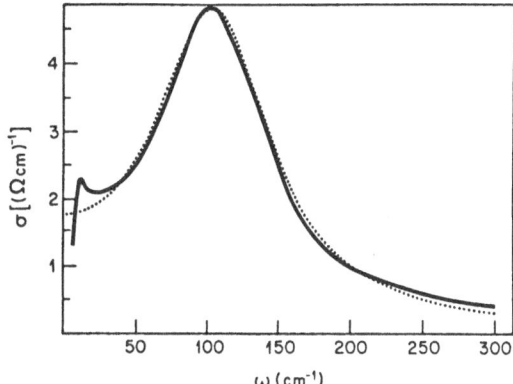

Fig. 7.9. Conductivity $\sigma(\omega)$ of α-AgI at $T = 453$ K. (—): derived from far-infrared reflectivity measurements; ($\cdots$): fit based on a memory function approach (7.23) [7.30]

cillatory regime the ion moves with an effective mass, an effective charge (the Szigeti charge, [Ref. 7.18, Sect. 4.3.1] and in an effective field, while in the diffusive regime it moves with the bare mass, the bare charge and in the external field. In this interpretation the shoulder observed in α-AgI near $15\,\mathrm{cm}^{-1}$ (Fig. 7.9) is a direct consequence of the transition from oscillation of the dressed particle to diffusion of the bare particle. Although it is possible that part of this structure results from polaronic effects, detailed studies to be discussed in Sect. 7.4 indicate that it is much more likely that it originates from disorder.

7.5 Disorder-Induced Infrared Absorption and Raman Scattering

The diffusive motion of ions in a SIC plays a crucial role when determing its low frequency response. At high frequencies, however, one expects that it is sufficient to consider static disorder and to neglect the diffusive motion altogether when evaluating response functions. The investigation of the dynamical properties of a SIC at high frequencies is therefore essentially a study of vibrations in a structurally disordered system.

There are several effects due to disorder. First of all, both the mobile ions and the cage ions may experience different forces when vibrating around their equilibrium positions. This gives rise to a fluctuation of the force constants which is due to different neighborhoods and results in a change of the vibrational density of states. Furthermore, due to a lack of translational periodicity, selection rules are broken. This may lead to disorder-induced infrared absorption and Raman scattering. The fluctuation of the force constants and the associated distribution of vibrational frequencies may also lead to very broad line widths in the inelastic neutron spectra.

The breakdown of the $q = 0$ selection rules for infrared absorption and Raman scattering can be accounted for qualitatively by considering the coherence length of the normal modes. In an ordered crystal, lattice dynamics shows us, that because of the periodicity of the lattice, the normal modes of vibration are plane waves with infinite extent. In the one-phonon infrared and Raman spectra only $q = 0$ optical modes can therefore be observed [Ref. 7.59, Chaps. 2, 3], which gives rise to the discrete set of lines in the spectra. In order to account for disorder in a qualitative way it may be assumed that the coherence length of the normal modes is short compared with the wavelength of light. As the coherence length shortens, the plane waves transform gradually into localized Einstein modes. It has been shown by *Shuker* and *Gammon* (SG) [7.60] that the assumption of short coherence lengths yields the breakdown of the usual $q = 0$ selection rules and allows the infrared- and light scattering process to occur essentially at all normal modes of the material.

Based on the SG model, *Galeener* and *Sen* [7.61] have developed an improved model for disorder-induced infrared absorption and Raman scattering. In contrast to SG the latter authors assume from the very beginning the complete absence of long-range order in atomic positions. In effect, the sample is treated as a giant molecule having no translational and rotational symmetry and consisting of N atoms, each vibrating harmonically about its equilibrium position. In both models it is assumed that the normal vibrations fall into bands $b = 1 \ldots B$ having similar microscopic motions, frequencies and optical coupling coefficients, e.g. stretching bands or bending bands. The final result for the frequency-dependent conductivity due to first-order infrared absorption can be written in the form [7.61]

$$\sigma(\omega) = \frac{\pi}{2} \sum_{b=1}^{B} D_b(\omega) \varrho_b(\omega) \ . \tag{7.24}$$

In this relation $\varrho_b(\omega)$ is the density of states per unit volume of the modes in band b, and

$$D_b(\omega) = \left| \sum_{l=1}^{3N} e_l^* u_l^b(\omega) \right|^2 \ . \tag{7.25}$$

In (7.25) e_l^* is an effective atomic charge and $u_l^b(\omega)$ denote the eigenvector components of the atoms for the type of modes contained in band b. Note that $D_b(\omega)$ is a measure for the square of the dipole moment induced by the displacements of the atoms in band b.

For the Raman response a similar expression can be derived. For first-order Raman scattering in the usual 90° scattering configuration [Ref. 7.59, Fig. 3.5], the result for anti-Stokes scattering is [7.61]

$$I_{\varrho\sigma}(\omega) = \text{const} \, (\omega_L + \omega)^4 \frac{\bar{n}(\omega)}{\omega} \sum_{b=1}^{B} C_b^{\varrho\sigma}(\omega) \varrho_b(\omega) \ , \tag{7.26}$$

where

$$C_b^{\varrho\sigma}(\omega) = \left| \sum_{l=1}^{3N} \alpha_{\varrho\sigma,l} u_l^b(\omega) \right|^2 \ . \tag{7.27}$$

Here, ω_L is the frequency of the incident light and ω is the vibrational frequency; $\bar{n}(\omega)$ is the mean occupation number of the mode with frequency ω. $\alpha_{\varrho\sigma}$ is the $\varrho\sigma$-component of the polarizability tensor ($\varrho, \sigma = x, y, z$) and $\alpha_{\varrho\sigma,l} = (\partial \alpha_{\varrho\sigma}/\partial u_l)_0$. A similar expression holds for Stokes scattering but with $(\omega_L + \omega)^4$ replaced by $(\omega_L - \omega)^4$ and $\bar{n}(\omega)$ replaced by $\bar{n}(\omega) + 1$ (compare with [Ref. 7.59, Eqs. (3.87, 88)]).

From (7.24, 26) it is evident that in general all vibrational modes of the disordered system will contribute to first-order infrared absorption and Raman scattering; both, $\sigma(\omega)$ and $I_{\varrho\sigma}(\omega)$ are related to a single set of densities $\varrho_b(\omega)$ of vibrational subband states. It should be pointed out that the eigenvector components $u_l^b(\omega)$ and hence the coefficients $D_b(\omega)$ and $C_b^{\varrho\sigma}(\omega)$ given by (7.25, 27) cannot be assumed to be independent of ω; in fact, they may vary strongly over a given subband [7.61]. In some cases the shape of a peak in the infrared or Raman spectrum of a disordered material may actually reflect the frequency dependence of the coupling coefficients $D_b(\omega)$ and $C_b(\omega)$ much more than that of the subband density of states $\varrho_b(\omega)$. Nevertheless, partition of the problem into subbands appears to be advantageous because some coefficients will be only weakly frequency dependent, and because the behaviour of $D_b(\omega)$ and $C_b(\omega)$ will be more easily understood than that of the coefficients $D(\omega)$ and $C(\omega)$ defined by

$$\begin{aligned} D(\omega)\varrho(\omega) &= \sum_b D_b(\omega)\varrho_b(\omega) \ , \\ C(\omega)\varrho(\omega) &= \sum_b C_b(\omega)\varrho_b(\omega) \ , \end{aligned} \tag{7.28}$$

where $\varrho(\omega) = \sum_b \varrho_b(\omega)$ is the total density of vibrational states. The evaluation of $\varrho(\omega)$, $D(\omega)$ and $C(\omega)$ is usually based on a computer calculation of the eigenfrequencies and eigenvectors of a large cluster of atoms subject to periodic boundary conditions. There are four ingredients that go into a calculation of the infrared conductivity

$$\sigma(\omega) \sim D(\omega)\varrho(\omega) \tag{7.29}$$

and the Raman spectra

$$I(\omega) \sim (\omega_L + \omega)^4 \frac{\bar{n}(\omega)}{\omega} C(\omega)\varrho(\omega) \ , \tag{7.30}$$

namely

a) a model for the equilibrium positions of the atoms of the disordered system,

b) a model for the interatomic forces,

c) expressions for the changes in the dipole moment [for $D(\omega)$] and electronic polarizability [for $C(\omega)$] associated with atomic displacements, and

d) a method for efficiently solving the harmonic lattice dynamical problem for systems with finite but large numbers of atoms.

In the simplest approximation $D(\omega)$ can be evaluated on the basis of a point charge model for the effective charges appearing in (7.25). With considerable less confidence one can assume atom or bond polarizabilities and calculate $C(\omega)$. Computer calculations of disorder-induced infrared absorption and Raman scattering of the SIC α-AgI based on the equation of motion method will be discussed in Sect. 7.6.1.

7.6 Phonons in α-AgI and Other SIC

7.6.1 Phonons in α-AgI

As an illustrative example we consider the high-frequency dynamical properties of α-AgI which is one of the best studied SIC. Figure 7.10 shows the temperature dependence of the ionic conductivity due to the Ag^+ ions measured in the microwave region at 22.0 GHz [7.62]; in contrast to older experiments the microwave conductivity is identical with the dc conductivity. Silver iodide is a normal ionic conductor below 420 K (β-AgI) and a SIC between 420 K and the melting temperature of 830 K (α-AgI). Note that according to Fig. 7.10 the ionic conductivity decreases at the transition from α-AgI to the ionic liquid.

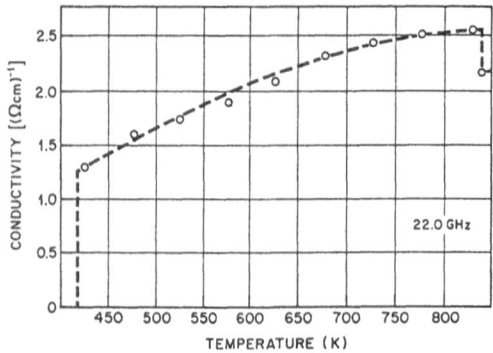

Fig. 7.10. Microwave conductivity of α-AgI versus temperature at 22.0 GHz. (- - -) indicates σ_{dc} [7.62]

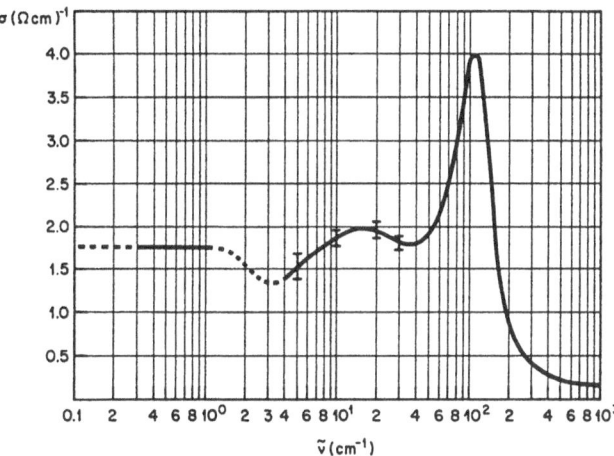

Fig. 7.11. Far-infrared and microwave conductivity of α-AgI at 250°C. $\tilde{\nu}>5\,\mathrm{cm}^{-1}$: [7.63]; $0.3\,\mathrm{cm}^{-1}<\tilde{\nu}<1.3\,\mathrm{cm}^{-1}$: [7.62]. (- - -): inter- and extrapolations. The vertical bars indicate the experimental errors

The crystal structure, phonon dispersion and density of states of β-AgI are illustrated in [Ref. 7.18, Figs. 3.7, 12, 13] as well as in [Ref. 7.18, Figs. 6.7a, 9]. The structure of the SIC α-AgI is shown in Fig. 7.2.

Figure 7.11 shows the far-infrared and microwave conductivity of α-AgI at 250°C. The far-infrared data have been obtained by *Brüesch* et al. [7.63], while the microwave data represent the new results obtained by *Funke* et al. [7.62a] and by *Roemer* et al. [7.62b]; in contrast to the older experiments which showed a considerable smaller microwave conductivity and a peak near $1\,\mathrm{cm}^{-1}$, the present new results demonstrate that α-AgI formed directly from the melt exhibits its dc conductivity σ_{dc} at microwave frequencies. The resonance in $\sigma(\omega)$ near $100\,\mathrm{cm}^{-1}$ is due to the TO mode in the disordered structure in which the silver ions vibrate against the iodine ions. This absorption is already observed in β-AgI [7.33] and survives the $\beta\rightarrow\alpha$ phase transition. The gross feature of $\sigma_{\mathrm{exp}}(\omega)$ as shown in Fig. 7.11 resembles $\sigma_{\mathrm{theor}}(\omega)$ as obtained from the model of the Brownian motion of a particle in a periodic potential (Fig. 7.8 with $k_{\mathrm{B}}T/U \cong 0.5$). In β-AgI $\sigma(\omega)$ is very small below about $50\,\mathrm{cm}^{-1}$ [7.63, 64] but heating the sample across the phase transition causes a drastic increase of $\sigma(\omega)$ and in the α-phase a pronounced shoulder is observed in the region around $15\,\mathrm{cm}^{-1}$. This shoulder cannot be explained by the model described in Sect. 7.3. Since it is the degree of disorder which changes abruptly at the $\beta\rightarrow\alpha$ phase transition, it is tempting to assume that the structure near $15\,\mathrm{cm}^{-1}$ is due to disorder-induced absorption as discussed in Sect. 7.5.

A more quantitative study of disorder-induced absorption in α-AgI has been given by *Alben* and *Burns* [7.65, 66]. These authors have calculated

185

the phonon density of states, the far-infrared conductivity $\sigma(\omega)$ and the Raman spectra $I(\omega)$ discussed in Sect. 7.5 by using the equation of motion method [7.67]. In this calculation the disordered structure is approximated by finite clusters of 1000–2000 atoms with periodic boundary conditions, and the dynamical properties are studied without diagonalizing the dynamical matrix. The clusters are rectangular, containing about 250 unit cells and the Ag^+ ions were distributed at random over the 12 tetrahedral sites, however, with the subsidary condition that when a tetrahedral site is occupied the four nearest-neighbour sites have to remain empty. The results are shown in Fig. 7.12. In Fig. 7.12 the Stokes scattering of the reduced Raman spectrum $I_{red}(\omega) = \omega I(\omega)/[1 + \bar{n}(\omega)]$ is plotted. If the modes observed by Raman scattering are simultaneously infrared active, as might be expected in disordered systems, it is possible to show that under certain conditions $I_{red}(\omega) \sim \sigma(\omega)$ [7.68–70]. Note the large peak in the density of states near 0.6 THz ($\sim 20\,cm^{-1}$) and the associated structures in $\sigma(\omega)$ and $I_{red}(\omega)$ which are to be expected from the relations (7.24, 26) discussed in Sect. 7.5.

Beyeler et al. [7.23] have obtained results similar to those shown in Fig. 7.12 using a different approach: In a first step a dynamical matrix for the primitive unit cell of a virtual crystal is constructed. The fluctuations of the force constants with respect to this virtual crystal are then introduced

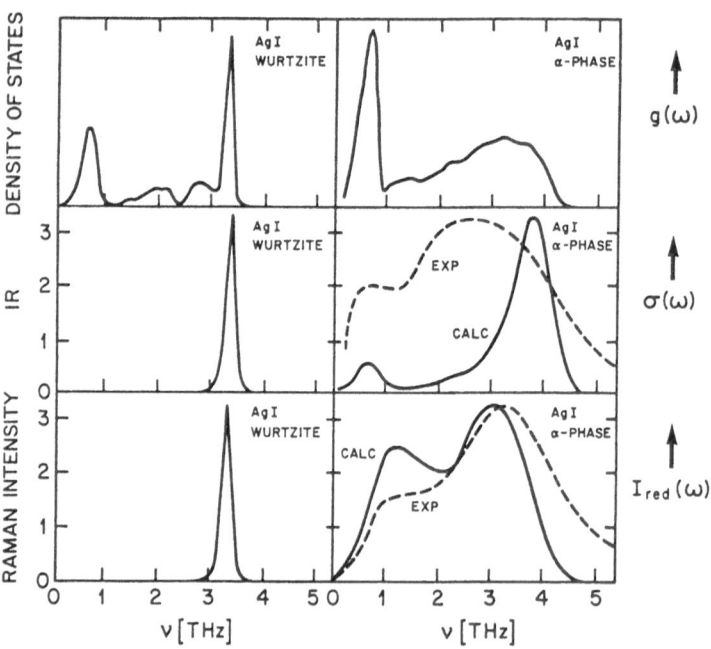

Fig. 7.12. Phonon density of states $g(\omega)$, far-infrared conductivity $\sigma(\omega)$ and reduced Raman intensity $I_{red}(\omega)$ for the β-phase and α-phase of AgI [7.65, 66]

and the problem is treated using a coherent-potential approximation (CPA) method. Both models, the computer calculation as well as the CPA method are able to account qualitatively for the observed far-infrared and Raman spectra and confirm that it is indeed the disorder which is mainly responsible for the low frequency structures in these spectra.

Inelastic neutron scattering experiments have been performed on a large single crystal of α-AgI by *Brüesch, Bührer* and *Smeets* [7.63]. Figure 7.13 shows the transverse (TA) and longitudinal (LA) acoustic phonon branches in the [110] direction observed at 300°C. The TA branch is very flat and a very large spread in energies about the mean branch is observed; this spread in energies is indicated by the vertical bars in Fig. 7.13. The energy at the N-point is only 2.5 meV with a width of as large as 4 meV. The mean TA branch is very similar to the TA branch in β-AgI [Ref. 7.59, Fig. 6.9] but in the latter case the widths are very much smaller. From a comparison of the dispersion curves of β-AgI and α-AgI and of the line widths of corresponding branches it is concluded that it is again mainly the disorder of the silver ions (rather than anharmonicity) which gives rise to the extremely large line widths of the TA modes in α-AgI.

Both, the computer calculation [7.65, 66] and the CPA method [7.23] take properly into account disorder but no information about the pattern of ionic displacements in the low-frequency modes is obtained. In order to obtain better insight into the ion dynamics responsible for the low-frequency modes observed around $20 \, cm^{-1}$ (Figs. 7.11–13), *Beyeler* et al. [7.23] and *Brüesch* et al. [7.63] have taken a different approach. A unit cube is considered which contains two Ag^+ and I^- ions only. There are four possible ways to distribute the Ag^+ ions over the available tetrahedral sites when nearest neighbour occupancies are excluded (Fig. 7.14). By repeating each

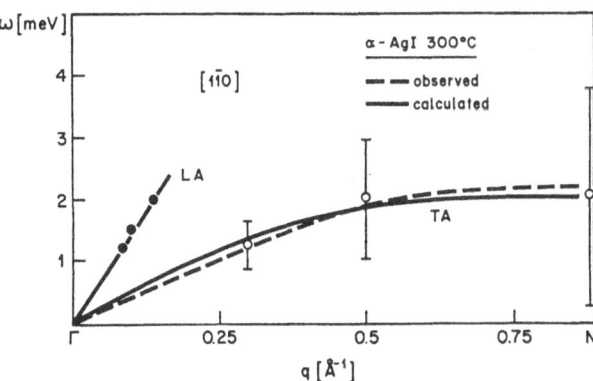

Fig. 7.13. Observed transverse (TA) and longitudinal (LA) acoustic phonon branches of an α-AgI single crystal at 300°C. The vertical bars indicate the spread in energies (linewidths) for the TA phonons propagating in the [1$\bar{1}$0]-direction. The calculated dispersion (—) is based on configuration 1 in Fig. 7.14 [7.63]

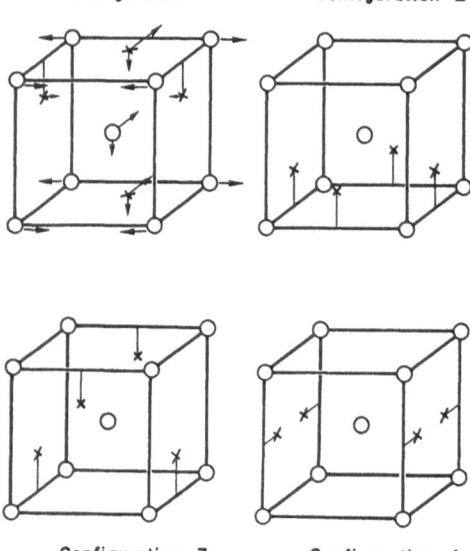

Configuration 1   Configuration 2

Configuration 3   Configuration 4

Fig. 7.14. Four different configurations of the silver ions in the cubic unit cube of α-AgI. All configurations with two silver ions in nearest-neighbor sites are excluded. The statistical weight of configuration 1 is four, that of configurations 2, 3 and 4 is only 1. (0) iodine ions, (x) occupied silver sites. The arrows in configuration 1 indicate the eigenvector of the [110] TA-zone-boundary mode (point N in Fig. 7.13) with a calculated frequency of $20\,\mathrm{cm}^{-1}$ [7.23, 63]

configuration periodically one obtains four different ordered "crystals" for which one can determine the eigenfrequencies and eigenvectors. Within this model disorder is taken into account only within this cube and the observed broad optical spectra (Figs. 7.11, 12) as well as the large linewidths of the inelastic neutron spectra (Fig. 7.13) result from a superposition of the phonon frequencies $\omega(\boldsymbol{q})$ of the four configurations. The force fields which are used include short-range as well as long-range Coulomb interactions and are taken from a fit to the observed phonon dispersion of β-AgI performed by *Buehrer* et al. [7.71] [Ref. 7.18, Fig. 3.12]. It is found that this force field not only reproduces the phonon dispersion but also the mean square amplitude of vibrations as determined from elastic neutron and X-ray scattering experiments. Thus one also expects the eigenvectors to be well approximated. Calculations for the four configurations show that the [1$\bar{1}$0] TA phonons have a particular flat dispersion curve for all four cases in agreement with the experiments shown in Fig. 7.13. This leads to a high density of states around $20\,\mathrm{cm}^{-1}$ in agreement with Fig. 7.12. The corresponding eigenmode at the zone boundary (point N in Fig. 7.13) for the configuration with the highest statistical weight (4 compared with 1 for the other configurations) is shown in Fig. 7.14. Inspection of this mode shows that the displacements of the ions are such that little or no changes occur in the nearest Ag-I and I-I distances, thereby minimizing the mutual repulsion between the ions. This kind of motion favours the jump of Ag^+ ions into neighbouring empty sites, and the mode illustrated in configuration 1 of Fig. 7.14 represents an approximate *reaction-coordinate* for such jumps (Sect. 7.7).

7.6.2 Phonons in Other SIC

β-Ag_3SI is a silver ion conductor with an ionic conductivity of about 10^{-2} $(\Omega\,cm)^{-1}$ at 295 K. It crystallizes in the real anti-perovskite structure with the S^{2-} ions at the center and each of the 3 Ag^+ ions distributed over 4 equivalent sites on the faces of the cubic unit cell. These 4 equivalent sites are close to the face center and separated by a small potential barrier Δ. The frequency dependent ionic conductivity $\sigma(\omega)$ has been determined from far-infrared reflectivity measurements by *Brüesch* et al. [7.72] and by *Gras* et al. [7.73]. The low-frequency part of $\sigma(\omega)$ can successfully be interpreted by a model of a Brownian particle (Ag^+ ion) in a double-well potential with barrier Δ [7.72]. This model gives an approximate description of the Ag^+ ions rattling between the fourfold equivalent sites on the faces of the cubic unit cell.

α-$RbAg_4I_5$ is probably the SIC with the highest ionic conductivity at room temperature, namely $\sigma_{dc} = 0.2$ $(\Omega\,cm)^{-1}$; this conductivity is entirely due to the Ag^+ ions. $\sigma(\omega)$ has been measured by *Eckold* and *Funke* [7.74] and is very similar to that of α-AgI shown in Fig. 7.11. In addition to the main resonance near $100\,cm^{-1}$, a pronounced structure near $20\,cm^{-1}$ is observed which originates from disorder-induced absorption. The Raman spectra of $RbAg_4I_5$ have been observed by *Delaney* and *Ushioda* [7.75] and by *Gallagher* and *Klein* [7.76] The reduced Raman spectra show features which are qualitatively similar to $\sigma(\omega)$ although the low frequency structures appear at different frequencies.

$AgCrS_2$ crystallizes in a layer structure. The most interesting aspect lies in the fact that the conducting silver ions are constrained to move in essentially two dimensional planes. Thus $AgCrS_2$ can be regarded as a quasi two-dimensional SIC. The phonons of $AgCrS_2$ have been studied by *Brüesch* et al. by means of far-infrared and inelastic neutron scattering techniques [7.77]. The most interesting feature of the phonon structure is the existence of very low-frequency acoustic and flat optic modes. The low-frequency TO mode at $q = 0$ shows a strong and anomalous temperature dependence down to 10 K. This temperature dependence can be explained by highly anharmonic motion of an Ag^+ ion in an effective potential. The shape of this potential has been determined from a fit of the model potential to the observed temperature dependence of $\omega_{TO}(q = 0)$. In the low-frequency modes the silver ions are strongly involved and vibrate parallel to the layers. These modes therefore contribute strongly to the mean-square displacements of the silver ions and to the reaction coordinates for ionic jumps.

$(C_5H_5NH)Ag_5I_6$, pyridinium pentasilver-hexaiodide, is a solid electrolyte with a very high conductivity of the Ag^+ ions, comparable to those of the best ionic conductors such as AgI and $RbAg_4I_5$. Far-infrared and incoherent neutron scattering experiments performed by *Brüesch* et al. [7.78] show the existence of very low-frequency optical modes around $20\,cm^{-1}$

which are of prime importance for the conduction mechanism of the Ag^+ ions.

α'-CuI stable between 407 and 465°C, is a SIC for Cu^+ ions. Raman spectra have been studied by *Burns* et al. [7.79,80]. Using a similar model as that applied to α-AgI (Sect. 7.6), the latter researches have calculated the density of states, infrared conductivity and Raman scattering and have shown that it is possible to explain the main features of the experimental results on the basis of disorder induced effects. It has also been confirmed that in contrast to the older measurements by *Funke* et al. [7.81] the microwave conductivity does not contain any structure and is the same as the dc conductivity [7.62].

$CaF_2, SrF_2, BaF_2, PbF_2$ crystallize in the fluorite structure. On heating, these compounds are found to exhibit broad specific heat anomalies at temperatures T_c well below the melting temperature T_m (for PbF_2: $T_c = 705\,K, T_m = 1158\,K$). The intrinsic defects most readily produced on heating fluorites are complex short-lived clusters composed of anion Frenkel pairs and neighbouring anions relaxed from their regular sites [7.99,100]. It is therefore assumed that the specific heat anomaly is due to the cooperative production of such defects giving rise to extensive disorder in the anion sublattice. These compounds are therefore anion conductors and their conductivity increases rapidly with increasing temperature approaching a value of about 10 $(\Omega\,cm)^{-1}$ at T_m.

Raman scattering measurements have been made on the fluorites as a function of temperature by *Elliot* et al. [7.82]. These crystals have one Raman active phonon with T_{2g} symmetry whose peak position and width change with temperature due to anharmonicity (Fig. 7.15a,b). At T_c and higher temperatures additional scattering develops on the low energy side of the T_{2g} phonon (Fig. 7.15c,d). The harmonic lattice dynamics of crystals with the fluorite structure is well known [7.83] and it is found that below T_c the position and shape of the Raman line can be described accurately by third- and fourth-order anharmonicity treated to the lowest order in perturbation theory (Fig. 7.15a,b). The additional low energy scattering at T_c and above is related to a broadened single-phonon density of states [7.82] and can be accounted for by a theory of defect-induced scattering (Fig. 7.15c,d).

Sodium β-alumina is the prototype of the β-alumina structure; if available in stoichiometric form, sodium β-alumina would have the formula $Na_2O:11Al_2O_3$. Its superionic properties are determined by departure from stoichiometry. The crystal consists of spinel blocks (Al_2O_3) which are separated by mirror planes containing sodium and oxygen ions. The sodium ions are relatively free to move in these planes, thus we are dealing again with a two-dimensional SIC as in $AgCrS_2$. Sodium β-alumina is an important solid electrolyte for technical applications in energy storage and conversion schemes. An interesting aspect lies in the fact that Na can be replaced by K, Rb and Ag which makes it particularly attracting for studying the dy-

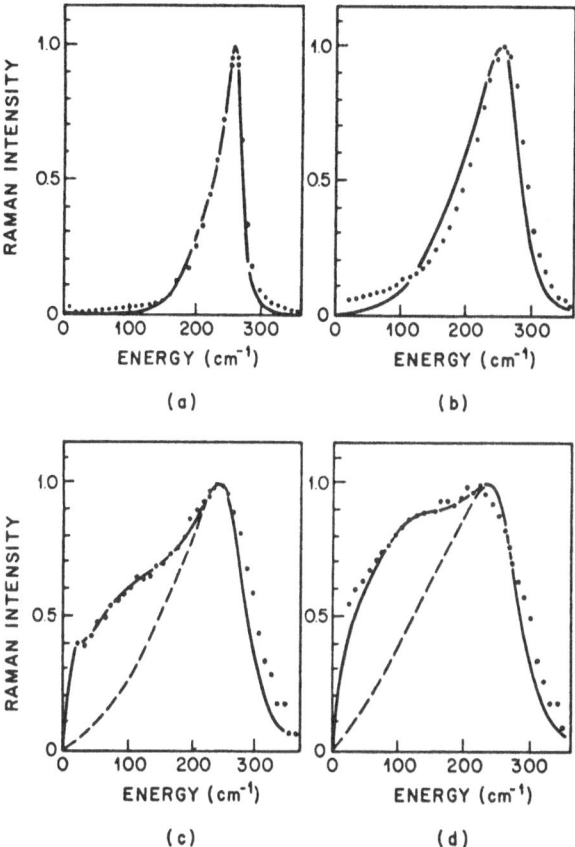

Fig. 7.15a–d. Measured (• • •) and calculated (—) reduced Raman intensities of PbF$_2$ at (a) 290 K, (b) 580 K, (c) 775 K and (d) 880 K ($T_c = 750$ K). The (- - -) in (c) and (d) represent the theoretical contribution of anharmonicity alone [7.82]

namical properties. The infrared and Raman spectra have been measured by several authors [7.33, 84–87] and reveal low-frequency modes which are closely related with the motion of the Na$^+$, K$^+$, Rb$^+$ and Ag$^+$ ions parallel to the conducting planes.

From the above considerations it is evident that disorder accounts for the main features observed in the far-infrared, Raman- and inelastic neutron scattering experiments of SIC. In all SIC considered above the existence of low-frequency modes is well established (Table 7.1).

The existence of such low-frequency modes appears to be a necessary but not a sufficient condition for SIC. Remember that the well-ordered β-AgI has also low-frequency modes but is a normal ionic conductor [Ref. 7.59, Fig. 6.9]. If, however, there is structural disorder with many available sites

Table 7.1. Regions of low frequency modes and mobile ions in some superionic conductors

SIC	Region of low-frequency structures $[\text{cm}^{-1}]$	Mobile ion
α-AgI	10–30	Ag^+
Ag_3SI	10–20,70	Ag^+
$RbAg_4I_5$	~20	Ag^+
$AgCrS_2$	~30	Ag^+
α'-CuI	~40	Cu^+
Na β-alumina	60	Na^+
Ag β-alumina	30	Ag^+
PbF_2 $(T>T_c)$	20–100	F^-

for the mobile ions, the low-frequency modes can be regarded as approximate *reaction coordinates* for ionic jumps.

7.7 Lattice Dynamical Calculation of Jump Frequencies

In Sect. 7.1 we have introduced a jump frequency

$$\nu_j = \nu_0\, e^{-U/k_B T} \tag{7.31}$$

for a particle in a potential with energy barrier U, ν_0 being the attempt frequency to overcome this barrier U (Fig. 7.3). The diffusion constant D is then proportional to $a_0^2 \nu_j$ (a_0 = jump distance) and the dc conductivity is given by (7.1) with $n \cong n_0$ for a SIC.

It is the purpose of this section to outline the evaluation of ν_j on the basis of lattice dynamical models. A calculation of jump frequencies in a SIC is evidently a complex task due to the presence of disorder and anharmonicity. In Sect. 7.7.1 we therefore illustrate the basic ideas for atomic jump processes in well-ordered crystals with a small defect concentration; here, we follow closely the work of *Flynn* [7.88,89]. In Sect. 7.7.2 we discuss possible applications of these results to SIC.

7.7.1 Dynamical Theory of Diffusion in Crystals with a Small Defect Concentration

Even for a crystal with a small defect concentration, $n \ll n_0$, atomic diffusion is an extremely complex phenomenon. Lying well outside the bounds of conventional harmonic lattice theory, its mechanisms are more grossly anharmonic than any likely fluctuations of the perfect lattice. Moreover, the elementary jumps always occur in a defective region of the lattice, where translational symmetry no longer prevails. In the present context we shall ignore this latter difficulty and assume that the perturbed lattice modes near a vacancy is sufficiently similar to unperturbed states. For the time being we shall even ignore anharmonicity because the simplifications introduced by considering the thermal motion to be harmonic is so extensive that we shall use it despite its obvious shortcomings. In calculating the jump frequency of an atom from one site to a neighboring energetically equivalent site we use in fact an approximation to the real anharmonic potential energy as illustrated in Fig. 7.16b.

To illustrate the main ideas consider an atom in a crystal which jumps from its site A to a neighbouring vacant site B (Fig. 7.16). The atom near A will be surrounded by a cage of other atoms; in the case of β-AgI and α-AgI this cage consists of 4 iodine ions which surround the silver ions tetrahedrally. If the Ag^+ ion leaves the cage through an edge of the tetrahedron, there are two iodine ions obstructing directly the escape of the Ag^+ ion (Fig. 7.16a); in the general case there will be a ring of barrier atoms which form the bottleneck for the escape. In the case of Fig. 7.16a the *reaction coordinate* $x(t)$ leading to a jump of the atom at site A into the vacant site B is a displacement in the x direction towards the saddle point S defined by the two iodine ions. It seems intuitively reasonable that short-range repulsions between the diffusing atom and certain of its cage atoms, the two iodine ions in Fig. 7.16a, dominate the dynamics of the jump process. These forces must therefore be emphasized in a correct selection of the reaction coordinate. Furthermore, the criterion for completion of a jump is closely tied to the position of a migrating atom with respect to its neighbors barring

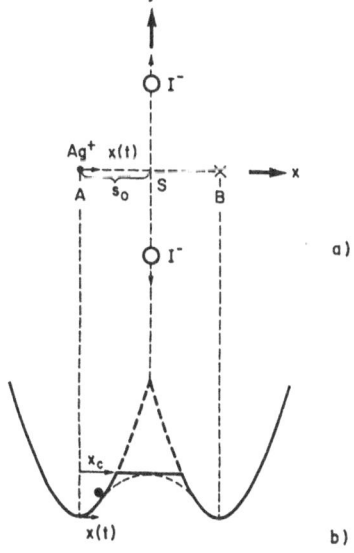

a)

b)

Fig. 7.16. Hypothetical potential energy of a jumping atom or ion as the reaction coordinate $x(t)$ varies. The harmonic potential and a potential cut-off at x_c so as to be consistent with the harmonic approximation are shown

the jump path. The neighbors exert strong repulsive forces on the jumping atom as it approaches the saddle point, and the jump will be completed only if the fluctuation is sufficiently strong to force the moving ion past its neighbors towards the vacancy. Evidently, the instantaneous *relative* position of the barrier atoms and the diffusing atom is of central interest in the diffusion process. As a first crude approximation to the dynamics of the jump process the following reaction coordinate is defined [7.89]:

$$x(t) = u_x(\mathbf{r}_d, t) - \frac{1}{Z} \sum_n u_x(\mathbf{r}_n, t) \ . \tag{7.32}$$

Here, $u_x(\mathbf{r}_d, t)$ is the displacement of the diffusing atom along the jump path which is taken to be the x direction, and the $u_x(\mathbf{r}_n, t)$ are the displacements of the Z barrier atoms along the jump path; $\mathbf{r}_d$ and $\mathbf{r}_n$ are the equilibrium positions of the diffusing atom and the barrier atoms, respectively. In Fig. 7.16a, $Z = 2$ and the jump path is parallel to the x axis. In effect, the quantity $Z^{-1} \sum u_x(\mathbf{r}_n, t)$ defines the moving saddle point as the center of gravity of these Z neighbors.

With (7.32) we lose immediately any manybody aspect of the jump process; attention is focused on the behaviour of a particular set of atoms crucial to the completion of a jump, and the configuration of all other atoms is ignored. This point will be further discusses in Sect. 7.7.2. Equation (7.32) is open to criticism on the grounds that, even inside the assumption that certain chosen atoms alone are important, a relevant part of their motion is still ignored; for it is obvious that the *lateral* motion of barrier atoms is essential to minimize the required fluctuation energy. This point has been emphasized in the theory by *Rice* [7.90]. In Fig. 7.16a this would correspond to an out-of-phase motion of the two iodine ions as indicated by two arrows. The simultaneous occurence of a displacement of the Ag^+ in the x direction *and* an out-of-phase motion of the iodine ions in the y direction favours the jump because a hole opens at the saddle point through which the ion can pass easily. In fact, a better reaction coordinate could be defined which takes into account the lateral motion of the barrier atoms, and a pure bending coordinate, that is, a change in the I-Ag-I bond angle [Ref. 7.18, Sect. 4.5] would probably be a good choice. However, it is likely that the effect of lateral motion of the barrier atoms enters into the theory only as an adjustment of the critical value x_c of the reaction coordinate x. x_c measures the type and size of fluctuation needed to effect a jump. In the following it is assumed that jumps occur when the reaction coordinate x exceeds the critical value x_c. The critical value x_c is not given by the present theory; it must be obtained from geometrical considerations of the lattice and from a comparison with experimental diffusion data.

The dynamical approach considered here treats the reaction coordinate as defined by (7.32) as a superposition of phonons in the harmonic crystal. Each phonon displaces a mobile ion towards the saddle point. Since the phonon phases are random the displacements may occasionally coincide in such a way that a jump occurs. The large local displacement that results from a jump is thus Fourier synthesized like a δ-function from the eigenvectors of the dynamical matrix; it is related to the average atomic displacements as a pistol shot is related to white-noise, by a phase coincidence among a broad spectrum of Fourier components.

Formulating these ideas more quantitatively we represent the local displacement $u(r,t)$ of the atom with equilibrium position r as a superposition of the normal modes of the system; in the present context we use a slightly different notation as usual (see for example [Ref. 7.18, Eq. (3.53)]) and write

$$u(r,t) = \sum_s u_s(r)e^{i\omega_s t} \ , \quad \text{where} \tag{7.33}$$

$$u_s(r) = e_s u_s^0 e^{iq \cdot r} \ . \tag{7.34}$$

Here, e_s is the polarization vector of the mode $s = (q,j)$, q is the wavevector, j the phonon branch and ω_s the eigenfrequency. Substituting $(7.33, 34)$ into (7.32) gives

$$x(t) = \sum_{s=1}^{N} x_s^0 e^{i\omega_s t} \ , \tag{7.35}$$

where N is the number of oscillators,

$$x_s^0 = u_{sx}(r_d) - \frac{1}{Z}\sum_{n=1}^{Z} u_{sx}(r_n) \ , \quad \text{and} \tag{7.36}$$

$$u_{sx}(r) = e_{sx} u_s^0 e^{iq \cdot r} \ , \tag{7.37}$$

e_{sx} being the component of the polarization vector e_s in the x direction, i.e. along the jump path (Fig. 7.16a). The reaction coordinate $x(t)$ as defined by (7.35) will fluctuate wildly as a function of time as illustrated schematically in Fig. 7.17 [7.101].

According to our assumption atomic jumps occur when the fluctuation in the reaction coordinate $x(t)$ given by (7.35) exceeds the critical value x_c. The criterion for a jump is that $x(t) - x_c$ has an "upzero" (i.e., swing through zero in an increasing sense). Upzero frequencies of sums such as (7.35) may be obtained from *Kac*'s relation [7.91]. *Kac*'s general relation may be applied to any harmonically vibrating system, including molecules, but for a relatively small number of vibrating particles the relation is rather complicated. If the number of oscillators is large, as in a crystal, an approximation to *Kac*'s relation due to *Slater* [7.92] may be used. When

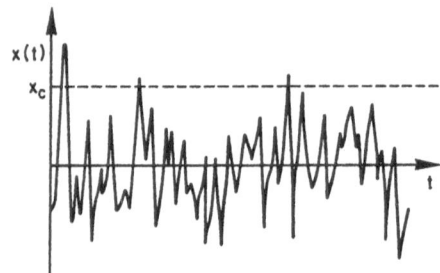

Fig. 7.17. Schematic illustration of the time dependence of the reaction coordinate $x(t)$. Atomic jumps occur if $x(t) > x_c$ [7.101].

$$x_c \ll \sum_{s=1}^{N} x_s^0 \; , \tag{7.38}$$

one may represent displacements by a Gaussian distribution, under which condition the upzero frequency or jump frequency is

$$\omega_j = \left[\frac{\sum\limits_s \omega_s^2 |x_s^0|^2}{\sum\limits_s |x_s^0|^2} \right]^{1/2} \exp\left(-x_c^2 / \sum_s |x_s^0|^2\right) \; . \tag{7.39}$$

The pre-exponential in (7.39) represents a mean attack frequency and can be identified with the attempt frequency $\omega_0 = 2\pi\nu_0$ in (7.31). The exponential in (7.39) gives the probability that the system shall be found at the critical displacement x_c. This probability increases with increasing temperature, since at high temperatures $|x_s^0|^2 \to k_B T$. Equation (7.39) is asymptotically exact when the x_s^0 are all equal and N is large. *Feit* [7.93] has shown that (7.39) is also exact when the x_s^0 are projections onto an arbitrary axis in configuration space of classical lattice vibrational amplitudes for thermal equilibrium. This is the case relevant for the present discussion.

The lattice potential energy associated with the diffusing atom as approximated by the present model, is shown in Fig. 7.16b. It is harmonic up to $x = x_c$ and for any displacement larger than x_c the jump proceeds to completion. By applying (7.39) to a suitable reaction coordinate of a mobile ion we may predict the jump frequency ω_j and hence the mobility μ given by (7.3a) in terms of a single parameter, x_c, which measures the size of fluctuation necessary to cause an atomic jump.

Flynn [7.88, 89] has developped a theory of the diffusion process in monoatomic cubic crystals on the basis of (7.39). To obtain ω_j it is necessary to evaluate the values of the amplitudes x_s^0. These may be obtained in principle from the phonon spectrum and using (7.36, 37). On the basis of the Debye approximation to the phonon spectrum [Ref. 7.18, Eq. (3.94)], the following expression, valid at high temperatures, is obtained for the jump frequency

$$\omega_j = \left(\frac{3}{5}\right)^{1/2} \omega_D \exp\left(-\frac{15m\delta^2}{2k_BT(3v_L^{-2} + v_{T_1}^{-2} + v_{T_2}^{-2})}\right) . \tag{7.40}$$

In this expression ω_D is the average Debye frequency obtained from specific heat data [Ref. 7.18, Eqs. (3.106, 107)] and v_L, v_{T_1} and v_{T_2} are sound velocities of longitudinal and transverse branches of the phonon spectrum expressed by $\omega_s = v_s q$. The sound velocities v_s (v_L, v_{T_1}, v_{T_2}) can be related to the elastic constants of the crystal [Ref. 7.18, Sect. 3.6, Problem 3.8.5]. m is the atomic mass and $\delta^2 = x_c^2/s_0^2$, where s_0 is the distance from the equilibrium position of the mobile ion to the saddle-point (Fig. 7.16). Comparison of the theory with experimental values of diffusion data and using experimental values of the elastic constants gives $\delta^2 = 0.104$ for fcc crystals and $\delta^2 = 0.067$ for bcc crystals [7.89]. Thus values of $x_c \cong 0.3s_0$ are sufficient to cause a diffusion jump.

An approximate elastic constant C, with which to represent the velocities in the exponent of (7.40) may be written as follows:

$$\frac{15}{C} = \frac{3}{C_{11}} + \frac{2}{C_{11} - C_{12}} + \frac{1}{C_{44}} , \tag{7.41}$$

where C_{11}, C_{12} and C_{44} are the elastic constants of a cubic crystal [Ref. 7.18, Problem 3.8.5]. The jump frequency may then be written in the form

$$\omega_j = \left(\frac{3}{5}\right)^{1/2} \omega_D \exp\left(-C v_a \delta^2/k_BT\right) . \tag{7.42}$$

Comparing this relation with (7.31) we obtain for the attempt frequency $\omega_0 = (3/5)^{1/2}\omega_D$ and for the energy barrier $U = C v_a \delta^2 = C v_a x_c^2/s_0^2$; thus the quantity $C v_a \delta^2$ simply represents the potential energy $(1/2)f x^2$ in the potential bowl a distance x_c along s_0 from the equilibrium point A (Fig. 7.16). A small elastic constant C and a small value of δ leads to a small force constant f and a small barrier U and hence to a large jump frequency.

The method outlined above for the calculation of ω_j for atoms in monoatomic lattices can be extended to crystals with a basis. A lattice dynamical calculation of the jump frequency of calcium ions in CaF_2 based on the reaction coordinate model has been performed by *Haridasan* et al. [7.94].

7.7.2 Jump Frequencies for SIC

Due to the structural disorder present in SIC the dynamical calculation of jump frequencies is certainly more complicated than in crystals with a small defect concentration. In a SIC it is, in general not possible to represent the reaction coordinate as a superposition of phonons, that is, the plane-wave approximation as expressed by (7.33, 34) is not applicable, unless simplifying assumptions are made (see below). Furthermore, as a consequence of disorder, each mobile ion is in general in a slightly different environment; even if the cage formed by the ions of the host lattice is the same, the different configurations of the neighboring mobile ions will lead to a distribution of restoring forces for the mobile ions and hence also to a certain distribution of jump frequencies. In the following we continue to assume that the short-range repulsions between a mobile ion and certain of its neighbour ions of the host lattice dominate the dynamics of the jump process. We therefore still use the definition (7.32) for $x(t)$ which implies that we neglect correlations between ionic jumps. The effects of correlations are not always negligible [7.102]; for topological reasons they are generally more important in low-dimensional SIC such as *Hollandite* [7.23, 103] where the interactions between the mobile ions may be stronger than the periodic potential of the host lattice, and diffusion is by no means a single-particle process. Confining ourselves to systems for which correlations may be neglected the jump frequency can, in principle, still be calculated on the basis of (7.39) with x_s^0 defined by (7.36). It should be noted that (7.39) yields the jump frequency when the frequency of fluctuation is considerably greater than the frequency of a jump. This condition is satisfied for most SIC.

In a simplified model, disorder may be treated very crudely as we have illustrated in the case of α-AgI (Fig. 7.14). In this model we have considered disorder of the Ag^+ ions only within a unit cube of the iodine sublattice; this leads to 4 different configurations shown in Fig. 7.14 and to 4 different ordered virtual crystals to which conventional lattice dynamics may be applied. For each crystal a suitable reaction coordinate may be chosen and each reaction coordinate may be represented in terms of phonons of the corresponding virtual crystal. In the case of α-AgI this procedure will lead to 4 slightly different jump frequencies and the actual jump frequency is a weighted average, weighted with the statistical multiplicity of the different configurations. This renders configuration 1 in Fig. 7.14 especially important because its statistical weight is 4 compared with 1 for the other three configurations. In fact the pattern of ionic displacements of configuration 1 as obtained from a lattice dynamical calculation [7.23, 63] can be regarded as a suitable reaction coordinate for jumps of Ag^+ ions to neighboring empty sites. This reaction coordinate is, however, more complicated than that defined by (7.32) because it contains the lateral motion, that is, the out-of-phase motion of the two iodine ions 1 and 2 which define the saddle-point, thereby minimizing the required fluctuation energy; in fact, the mean energy of this mode (energy at the N-point in Fig. 7.13) is only 2.5 meV.

In a more realistic treatment of disorder the disordered structure may be approximated by a finite cluster containing a relatively large number of ions, which is subjected to periodic boundary conditions. If the number of ions in the cluster is not too large, the use of rapid digital computers still allows the determination of eigenfrequencies and eigenvectors by direct diagonalization of the force constant matrix. In the case of α-AgI the disordered structure might be approximated by distributing the Ag^+ ions over the possible sites of 8 or perhaps even 27 unit cubes of the iodine sublattice. On this basis, suitably chosen reaction coordinates can again be represented as a superposition of a large number of eigen-modes of the finite cluster, and the jump frequencies may be evaluated on the basis of the Kac-Slater relation (7.39). Independent on the specific model used to calculate ω_j, the expression (7.39) shows that ω_j is large in SIC due to the high density of low frequency modes with large vibrational amplitudes.

Finally, it is interesting to note that the dynamical theory for atomic or ionic diffusion in solids presented here has common features with Lindemann's relation for melting discussed in the next chapter. In both theories a critical atomic displacement is introduced which measures the size of fluctuation necessary to trigger the process in question: an atom or ion jumps to a vacant site if the reaction coordinate x reaches a critical value x_c, and according to Lindemann the solid melts when the displacements of atoms due to thermal fluctutations reach a critical fraction of the interactomic distance. In fact, the highly mobile state of one type of ions in a SIC can be regarded as a premature melting of a sublattice in the rigid framework constituted by the ions of the host lattice. Similar concepts are used in the reaction rate theory of molecules in gases which describes dissociation processes or structural changes of the colliding molecules [7.92,95]. Finally, the absolute reaction rate theory developed by *Glasstone, Laidler* and *Eyring* plays a fundamental role in heterogeneous catalysis [7.105,106].

8. Melting (By L. Pietronero)

The general features of the melting transition for three-dimensional simple solids are critically discussed. In particular, the relations between melting and the instabilities of anharmonic dynamics are treated in some detail. The possibility of surface melting is also considered.

8.1 Nature of the Melting Transition

Despite the fact that the solid-liquid transition is one of the most familiar, the microscopic knowledge of this phenomenon is still rather scarce [8.1, 2]. We briefly discuss here the main questions connected with this transition especially in relation to the instabilities of the solid phase due to anharmonicity.

Since melting for simple three-dimensional systems is a first-order transition and does not show critical behavior, the first approach is to try to compute the free energies of the solid and liquid phases and define the melting point as the point where these two functions cross (T_M in Fig. 8.1). If we write

$$F_S = U_S(T) - TS_S(T) \ ,$$
$$F_L = U_L(T) - TS_L(T) \tag{8.1}$$

for the free energies (F), energy (U) and entropy (S) of the two phases, the melting temperature T_M is then given by

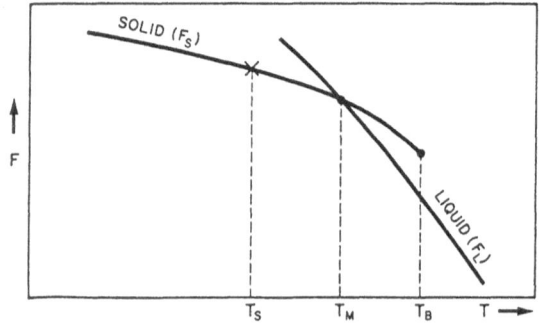

Fig. 8.1. Schematic view of the solid and liquid free energies (at fixed pressure) as functions of temperature. T_M refers to the real melting temperature while T_B represents the intrinsic instability temperature for the bulk of a solid. The free energy F_S between T_M and T_B refers to metastable states. T_S refers to surface melting

$$T_M = \frac{\Delta U_M}{\Delta S_M} \quad \text{with} \tag{8.2}$$

$$\Delta U_M = U_L(T_M) - U_S(T_M) \ ,$$
$$\Delta S_M = S_L(T_M) - S_S(T_M) \ . \tag{8.3}$$

A remarkable experimental fact is that, for a fixed volume, one finds in general [8.3, 4]

$$\Delta S_M \cong k_B \ln 2 \tag{8.4}$$

but it is hard to draw any conclusion from this observation.

The problem with this approach lies mainly in the difficulty of evaluating the free energy of the liquid phase [8.1]. But even if this would be possible important questions would still remain open. From this point of view in fact the Lindemann relation [8.5, 6] (that we discuss in the following) does not find a natural explanation. In addition, the absence of superheating of the solid phase [8.2] does not find any explanation from this point of view. Experiments show in fact that, while it is easily possible to supercool a liquid and reach the metastable states of F_L at the left of T_M (Fig. 8.1), it is not possible to superheat a solid and follow F_S on the right of T_M [8.2]. It is hard to see a reason for the free energy of the solid phase to loose the local stability just at T_M.

From the point of view of Landau's theory of phase transitions it is possible to conclude that, if the fluctuations around the transition are not too large, the transition should be first order because of symmetry reasons [8.7]. The Landau theory is based on an expansion of the free energy near the transition in powers of the average value of the order parameter, neglecting its fluctuations. Since large fluctuations do not seem to be present near the melting point (in simple 3-d systems) the Landau approach and its conclusions are probably correct. It is difficult however to go beyond qualitative arguments because of the difficulty of defining a proper order parameter for this transition. It is likely that several order parameters should be included in the theory. This is for example the case of a recent interesting approach to the freezing problem [8.8]. The method is based on the analysis of the instabilities of the homogeneous liquid phase with respect to a phase where density waves are present with wavevectors corresponding to the reciprocal lattice vectors of the solid.

Since here we are primarily concerned with the relations between melting and lattice dynamics we consider the problem from the complementary point of view: the instabilities of the solid phase. Historically one of the first approaches was to look at the melting transition as a lattice gas order-disorder transition [8.3]. Not much can be learned from this model that implies the existence of critical behavior never observed in melting. Several models have then been proposed that consider melting as an instability of the solid phase due to generation of defects or dislocations [8.9–11]. These approaches cannot be related directly to the real melting transition but should be more considered as models for the intrinsic instability of the solid phase (T_B in Fig. 8.1). The main problem is that these methods generally predict strong precursor effects (generation of many defect or dislocations before melting) that are not observed experimentally. The transition from solid to liquid in simple monoatomic systems is in general abrupt without appreciable precursor effects in terms of density of defects and dislocations or changes of the phonon frequencies and of the elastic properties [8.12]. The situation is different for some two dimensional systems for which the dislocation approach seems to be rather more appropriate [8.11].

In summary, the general facts one would like to understand about the melting transition in simple systems are:

(i) The absence of superheating of the solid phase;

(ii) The abrupt behavior of the transition without appreciable precursor effects. In addition one should also keep in mind the validity of the Lindemann relation that we discuss in the next section.

8.2 Lindemann Relation

In this and the following sections we discuss the features of melting that can be related to lattice dynamics and anharmonicity. The first argument in this direction is due to *Lindemann* [8.5,6]. He conjectured that melting should occur when the displacements u of the atoms due to thermal fluctuations reach a given fraction (C^*) of the interatomic distance d. This can be written as

$$\frac{\langle u^2 \rangle^{1/2}_{T_M}}{d} \equiv C^* . \tag{8.5}$$

Equation (8.5) represents what we may call today a scaling relation and we are now going to exploit its consequences.

We can easily evaluate the fluctuations of the atomic displacements within harmonic lattice dynamics ([8.6] and [Ref. 8.13, pp. 156–184]). This is certainly incorrect if one tries to make a theory of lattice instability because, as we shall see in the next section, anharmonicity has to be included. Nevertheless, since we know from experiments that phonon frequencies remain essentially unchanged up to melting, the use of harmonic dynamics to study the consequences of an *empirical conjecture* like (8.5) may be acceptable. We can then write the atomic displacement $u(l)$ in terms of creation and annihilation operators; for a Bravais lattice we have [Ref. 8.22, Appendix H]

$$u(l) = \sum_s \left(\frac{\hbar}{2Nm\omega_s}\right)^{1/2} (a^+_{-s} + a_s)e^{iq \cdot r(l)} , \qquad (8.6)$$

where q is the phonon wavevector, $s = (q, j)$ and j is the branch index. This gives

$$\langle u^2(l) \rangle = \sum_s \frac{\hbar}{2Nm\omega_s} \langle |a^+_{-s} + a_s|^2 \rangle . \qquad (8.7)$$

Since we are interested in evaluating (8.7) at the melting temperature we can take the classical limit $(k_B T_M \gg \hbar\omega)$ and write

$$\langle u^2 \rangle = \sum_s \frac{\hbar}{2Nm\omega_s}(2\bar{n}_s + 1) \cong \frac{k_B T}{Nm} \sum_s \omega_s^{-2} . \qquad (8.8)$$

Treating the phonons as Einstein modes with frequency corresponding to an average Einstein temperature θ_E gives then simply

$$\langle u^2 \rangle_{T_M} \cong \frac{3\hbar^2 T_M}{mk_B\theta_E^2} . \qquad (8.9)$$

For high temperature the assumption of Einstein modes is reasonable [8.6, 13]. However, it is also easy to evaluate (8.8) with a Debye model for the phonon spectrum. Neglecting the frequency difference between longitudinal and transverse modes we have [8.13]

$$\omega_j(q) = vq \quad \text{and} \qquad (8.10)$$

$$\langle u^2 \rangle_{T_M} = \frac{3k_B T_M}{mNv^2} \sum_q \frac{1}{q^2} . \qquad (8.11)$$

It is simple to use a continuum spherical integration to evaluate the sum over q in (8.11)

$$\sum_q \frac{1}{q^2} \cong N \frac{d^3}{(2\pi)^3} \int_0^{q^*} 4\pi \, dq . \qquad (8.12)$$

Caution has to be paid in the choice of q^*. For a cubic system the maximum value of each of the components of q is $q_{max} = \pi/d$. Here however we integrate over a sphere so we have to choose q^* in such a way as to satisfy the normalization condition

$$\frac{d^3}{(2\pi)^3} \int_0^{q^*} 4\pi q^2 \, dq = \frac{1}{N} \sum_q = 1 . \qquad (8.13)$$

This gives

$$q^{*3} = \frac{6\pi^2}{d^3} .$$
(8.14)

The Debye temperature θ_D is defined by the relation

$$k_B\theta_D = \hbar\omega(q^*) = \hbar v q^*$$
(8.15)

that gives finally

$$\langle u^2 \rangle_{T_M} = \frac{9\hbar^2 T_M}{mk_B\theta_D^2} .$$
(8.16)

This is the same relation, apart from a prefactor, as the one obtained with Einstein modes, (8.9).

Since not all structures are simple cubic we should rather use in (8.5) $d^* = v_a^{1/3}$ where v_a is the atomic volume. We obtain then

$$C^{*2} = \frac{9\hbar^2 T_M}{mk_B d^{*2}\theta_D^2} ,$$
(8.17)

a relation that is rather well satisfied for systems belonging to the same class with values of the constant order of $C^{*2} \cong 0.02$ [8.6].

The validity of this empirical relation strongly suggests the existence of a connection between the relative amplitude of the atomic displacements and the melting transition. It is interesting to note that an analogous empirical relation can also be found for the structural properties of fluids just before solidification. It has been observed in fact that the peak value of the structure factor at the first (largest) peak as a function of q is $S(q_m) \cong 2.85$ for all classical fluids at the freezing temperature [8.8].

8.3 Anharmonicity and Instability of the Solid Phase

In this section we discuss the instability of the solid phase when anharmonicity is introduced. We will show how it is possible to actually derive the Lindemann relation that was introduced as an empirical conjecture in the previous section. This approach of course, does not correspond to real melting (T_M in Fig. 8.1) but rather to the bulk instability temperature (T_B in Fig. 8.1).

Since we are interested in the ultimate anharmonic effect, the breakdown of the solid phase, it is not possible to use perturbative approaches to describe anharmonicity. The study of anharmonicity beyond perturbation theory is only possible within drastic and uncontrolled approximations. These are for example the quasi-harmonic (QH) and the self-consistent harmonic (SCH) approximations [Ref. 8.13, pp. 156–184]. The QH method has been used to study the instability of the quasi-harmonic free energy as a function of the lattice constants, provided a realistic pair potential between atoms is given [8.14]. Here we describe instead a simple model system within the SCH method. This approach is more useful to describe in a simple analytical form the general features of these instabilities.

Consider for simplicity a chain of atoms and let u_n be the component of the atomic displacements in the direction of the chain. The harmonic equations of motion are $(m = 1)$

$$\ddot{u}_n + K(n, n-1)(u_n - u_{n-1}) + K(n, n+1)(u_n - u_{n+1}) = 0 \qquad (8.18)$$

where the harmonic force constants are defined by

$$K(n, n') = \frac{\partial^2 V(n, n')}{du_n^2}\bigg|_0 \qquad (8.19)$$

and $V(n, n')$ is the potential between nearest atoms. This potential can in general be written as a Fourier series. Here we assume for simplicity that one term of the series dominates so we have just a sinusoidal potential

$$V(n, n') = \sum_q V_q \exp\left[iq(u_n - u_{n'})\right] \sim V_{q0} \exp\left[iq_0(u_n - u_{n'})\right] . \qquad (8.20)$$

The harmonic problem is then defined by (8.18, 19) and it is correct in the low temperature limit. Raising the temperature the nonlinearity of the potential, (8.20), is expected to produce strong deviations from the harmonic approximation. A way to introduce these effects is the SCH approximation ([Ref. 8.13, pp. 156–184] and [8.15, 16]). This method introduces an effective temperature dependent force constant $\tilde{K}(n, n')$ that is obtained by thermal averaging the second gradient of the potential. Selfconsistency is introduced by the fact that the thermal average is performed with an harmonic Hamiltonian whose force constants are again $\tilde{K}(n, n')$. We have [8.16]

$$\tilde{K}(n, n') = \left\langle \frac{\partial^2 V(n, n')}{\partial u_n^2} \right\rangle = -q_0^2 V_{q0} \langle \exp\left[iq_0(u_n - u_{n'})\right] \rangle$$
$$= -q_0^2 V_{q0} \exp\left(-\tfrac{1}{2} q_0^2 \langle |u_n - u_{n'}|^2 \rangle\right) . \qquad (8.21a)$$

Neglecting the cross terms in the exponent[1] we can then simply write

$$\tilde{K}(n, n') = K_0 \exp\left[-\lambda(\langle u_n^2 \rangle + \langle u_{n'}^2 \rangle)\right] \qquad (8.21b)$$

---

[1] This corresponds to Einstein oscillators whose force constants are affected by the thermal fluctuations of nearest neighbours.

where $K_0 = -q_0^2 V_{q0}$ is the harmonic (low temperature) force constant and λ is a factor depending only on the geometry of the system. For a homogeneous solid $\langle u_n^2 \rangle = \langle u_{n'}^2 \rangle$ and neglecting the dispersion in (8.18) (Einstein approximation) we obtain

$$\ddot{u}_n + 2\tilde{K}u_n = 0 , \qquad (8.22)$$

$$\tilde{K} = K_0 e^{-2\lambda\langle u_n^2 \rangle} . \qquad (8.23)$$

Equation (8.22) and the equipartition principle gives then

$$\tfrac{1}{2}k_B T = \tfrac{1}{2}2\tilde{K}\langle u_n^2 \rangle \qquad (8.24)$$

and therefore a selfconsistent relation is obtained for $\langle u_n^2 \rangle$ [8.16, 17]

$$\langle u_n^2 \rangle = \frac{k_B T}{2K_0}e^{2\lambda\langle u_n^2 \rangle} , \qquad (8.25)$$

or, introducing the dimensionless quantities

$$y = 2\lambda\langle u_n^2 \rangle , \qquad (8.26)$$

$$\tau = \frac{k_B T}{K_0}\lambda , \qquad (8.27)$$

$$y = \tau e^y . \qquad (8.28)$$

The values of y that satisfy (8.28) are easy to find graphically once τ is given (Fig. 8.2). At low temperature $\exp(y) \sim 1$ and we recover the harmonic solution $y = \tau$. Raising the temperature we obtain deviations

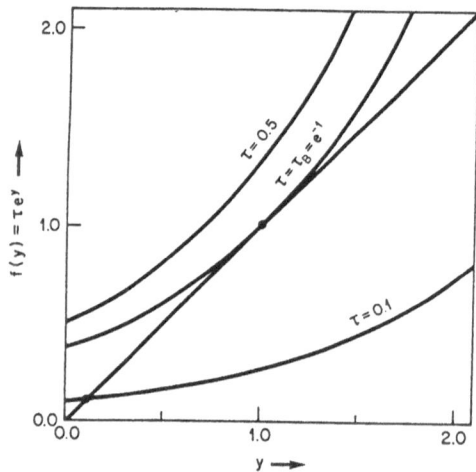

Fig. 8.2. Graphical solution of (8.28). At low temperature the harmonic solution is recovered. By raising τ anharmonic effects become more and more important until, above $\tau_B = e^{-1} = 0.368$ no more real solution exists

from this behavior until no more real solution can be found. It is easy to see that the maximum value of τ for which a real solution of (8.25) exists is

$$\tau_B = e^{-1} \cong 0.368 \tag{8.29}$$

with corresponding values

$$y_B = 1 \quad \text{and} \tag{8.30}$$

$$\tilde{K}_B = K_0 e^{-1} \ . \tag{8.31}$$

This means that above the temperature τ_B the solid phase with finite mean square displacements of the atoms is no longer stable because there is no longer a real solution to the selfconsistent equation that describes its dynamics. As previously mentioned this temperature should not be interpreted as the real melting temperature but rather as the bulk instability temperature T_B of Fig. 8.1. This instability is consistent with the Lindemann relation because (8.30) implies that the mean square displacements at T_M are

$$\langle u_n^2 \rangle_{T_M} = 1/2\lambda \ . \tag{8.32}$$

They depend therefore only on the geometrical parameter λ and *not* on the low temperature force constant K_0. The force constant decreases but it remains finite at the instability, (8.31), corresponding to a softening of phonon frequency of about 40% with respect to the low temperature value. In reality phonon softening at melting is actually much less then 40%. This fact is consistent witht the idea that our calculation refers to the intrinsic instability of the metastable solid phase T_B that should be rather higher than T_M. The question is then why this metastable region of the solid phase cannot be reached experimentally while supercooling of the liquid phase is instead easily possible.

A possible solution of this problem may arise from the fact that the surface of a solid seems to develop instabilities and becomes quasi-liquid at a temperature appreciably lower than the melting temperature [8.16]. This implies that the surfaces of a solid are already intrinsically unstable below T_M, providing unavoidable nucleation centers for the liquid phase and preventing therefore the possibility of overheating the solid phase.

One can easily generalize the model studied in this section to consider a semi-infinite chain with a free surface [8.16]. The complication this introduces is that the system is not any more homogeneous and the mean square displacements of the atoms are different for different distances from the free surface. The solution of the chain of non linear equations is then represented by the set of values $\{\langle u_n^2 \rangle; n = 1, 2, \ldots; N\}$ that, for atoms far from the surface, should converge to the bulk solution. It has been shown that the maximum stability temperature for the system with a free surface is $\tau_S = 0.270$ [8.16], about 30% lower than the bulk instability temperature $\tau_B = e^{-1} = 0.368$. This suggests therefore that, if real melting temperatures scale like these instabilities, there should be a temperature T_S lower then T_M at which the free surface of a simple solid becomes quasi liquid. Experiments and computer simulations are rather consistent with this result [8.17–21]. In fact, a recent experimental study on lead strongly supports such a picture [8.23].

An interesting consequence of this conclusion is that the reason that prevents the overheating of a solid is the presence of its intrinsically unstable surfaces. But if in some way this effect could be eliminated, then it should become possible to overheat the bulk of a solid. One way to "eliminate the surfaces" is to use periodic boundary conditions in computer simulations [8.20]. From the point of view of experiments one can use methods to heat only the inside of a solid leaving the surface cold [8.21]. The results obtained by these two methods confirm the possibility of overheating the bulk of a solid along the metastable free energy F_S between T_M and T_B (Fig. 8.1) and give therefore support to the surface instability picture.

Appendix: Constants and Units

General Physical Constants

Gas constant	$R = 8.31 \times 10^7 \, \text{erg mole}^{-1} \text{K}^{-1}$
Loschmidt's number	$N_L = 6.0225 \times 10^{23} \text{mole}^{-1}$
Speed of light in vacuum	$c = 2.998 \times 10^{10} \text{cm s}^{-1}$

Atomic Constants

Boltzmann's constant	$k_B = 1.381 \times 10^{-16} \text{erg K}^{-1}$
Planck's constant	$h = 6.626 \times 10^{-27} \text{erg } s$
Planck's constant $/2\pi$	$\hbar = 1.054 \times 10^{-27} \text{erg } s$
Electron rest mass	$m = 0.911 \times 10^{-27} \text{g}$
Proton rest mass	$M_p = 1.6725 \times 10^{-24} \text{g}$
Neutron rest mass	$M_n = 1.6747 \times 10^{-24} \text{g}$
Proton mass/electron mass	$M_p/m = 1836$
Atomic mass unit ($\equiv 1/16$ mass of O^{16})	$1 \, \text{amu} = 1.657 \times 10^{-24} \text{g}$
Charge on electron	$e = 4.803 \times 10^{-10} \text{esu}$
	$= 1.602 \times 10^{-19} \text{C}$
Bohr radius of the ground state of hydrogen	$a_0 = \hbar^2/me^2 = 0.529 \, \text{Å}$
Rydberg's constant	$1.09737 \times 10^5 \text{cm}^{-1}$

Length

Angström	$1.\text{Å} = 10^{-8}$ cm
Micron	$1 \, m\mu = 10^{-4} \text{cm}$

Energy conversion table

E	erg	J	cal	eV
1 erg =	1	10^{-7}	2.3892×10^{-8}	6.242×10^{11}
1 Joule =	10^7	1	2.3892×10^{-1}	6.242×10^{18}
1 cal =	4.1855×10^7	4.1855	1	2.613×10^{19}
1 eV =	1.602×10^{-12}	1.602×10^{-19}	3.827×10^{-20}	1

From $E = hc\bar{\nu} = k_B T$: $1 \, \text{eV} \hat{=} 8066 \, \text{cm}^{-1} \hat{=} 11605$ K

$1 \, \text{cm}^{-1} \hat{=} 1.438$ K; $1 \, \text{K} \hat{=} 0.695 \, \text{cm}^{-1}$

($\hat{=}$ means: corresponds to).

Force, Force Constants and Elastic Constants

Newton	$1\,N = 1\,kg\,ms^{-2} = 10^5\,dyn$
dyn	$1\,dyn = 1\,g\,cm\,s^{-2}$
mdyn	$1\,mdyn = 10^{-3}\,dyn$
Force constants	in dyn/cm or mdyn/Å
Elastic constants	in dyn/cm^2

Energy E

erg	$1\,erg = 1\,dyn\,cm$
Joule	$1\,J = Nm = 1\,W\,s$

Frequency ν

Hertz	$1\,Hz = 1\,s^{-1}$
Megahertz	$1\,MHz = 10^6\,Hz$
Gigahertz	$1\,GHz = 10^9\,Hz$
Terahertz	$1\,THz = 10^{12}\,Hz$

Wave Number $\tilde{\nu}$

Number of wavelengths λ per cm
$\tilde{\nu} = 1/\lambda = \nu/c$, in units of cm^{-1}
$(1\,THz \hat{=} 33.3\,cm^{-1})$

References

Chapter 1

1.1   P. Brüesch: *Phonons: Theory and Experiments I*, Springer Ser. Solid-State Sci.,
        Vol. 34 (Springer, Berlin, Heidelberg 1982)
1.2   P. Brüesch: *Phonons: Theory and Experiments II*, Springer Ser. Solid-State Sci.,
        Vol. 65 (Springer, Berlin, Heidelberg 1986)
1.3   J.M. Ziman: *Electrons and Phonons* (Oxford U. Press, London 1960) p. 357
1.4   J.M. Ziman: *Principles of the Theory of Solids* (Cambridge U. Press, Cambridge
        1964) p. 188
1.5   S.K. Sinha: In *Dynamical Properties of Solids*, Vol. 3, ed. by G.K. Horton and
        A.A. Maradudin (North-Holland, Amsterdam 1980) p. 1
1.6   P.B. Allen: In [Ref. 1.5, p. 95]
1.7   B. Lüthi: In [Ref. 1.5, p. 245]
1.8   P. Brüesch, F. D'Ambrogio: Phys. Stat. Sol. (b) **50**, 513 (1972)
1.9   T. Koehler: In *Dynamical Properties of Solids*, Vol. 2, ed. by G.K. Horton and
        A.A. Maradudin (North-Holland, Amsterdam 1975) p. 1

Chapter 2

2.1   W.P. Manson: J. Acoust. Soc. Am **70**, 1561 (1981)
2.2   W. Voigt: *Lehrbuch der Kristallphysik* (Teubner, Leipzig 1910)
2.3   J.F. Nye: *Physical Properties of Crystals* (Clarendon, Oxford 1957) p. 110
2.4   W.G. Cady: *Piezoelectricity* (Dover, New York 1964)
2.5   W.P. Warren: *Piezoelectric Crystals and their Applications to Ultrasonics* (Van
        Nostrand, Toronto 1950)
2.6   P. Brüesch: *Phonons: Theory and Experiments I*, Springer Ser. Solid-State Sci.,
        Vol. 34 (Springer, Berlin, Heidelberg 1982)
2.7   E. Gerlach, P. Grosse (Eds.): *The Physics of Selenium and Tellurium*, Springer
        Ser. Solid-State Sci., Vol. 13 (Springer, Berlin, Heidelberg 1979)
2.8   W.Ch. Cooper (Ed.): *The Physics of Selenium and Tellurium* (Pergamon, Oxford
        1969)
2.9   D. Royer, E. Dieulesaint: J. Appl. Phys. **50**, 4042 (1979)
2.10   W.D. Teuchert, R. Geick, G. Landwehr, H. Wendel, W. Weber: J. Phys. C **8**, 3725
        (1975)
2.11   H. Wendel, W. Weber, W.D. Teuchert: J. Phys. C **8**, 3737 (1975)
2.12   I. Chen, R. Zallen: Phys. Rev. **173**, 833 (1968)
2.13   R. Geick, U. Schröder, J. Stuke: Phys. Stat. Sol. **24**, 99 (1967)
2.14   P. Grosse, M. Lutz, W. Richter: Solid State Commun. **5**, 99 (1967)
2.15   G. Lucovsky, R.C. Keezer, E. Burstein: Solid State Commun. **5**, 439 (1967)
2.16a   R.G. Kepler, R.A. Anderson: CRC Crit. Rev. Solid State and Mat. Sci. (USA)
        9, 399 (1980)
2.16b   B. Jaffe, W.R. Cook, H. Jafe: *Piezoelectric Ceramics* (Academic, London 1971)
2.17   M. Born, K. Huang: *Dynamical Theory of Crystal Lattices* (Oxford U. Press,
        London 1954) p. 262
2.18   A.A. Maradudin, E.W. Montroll, G.H. Weiss, I.P. Ipatova: *Theory of Lattice Dy-
        namics in the Harmonic Approximation*, Solid State Physics, Suppl. 3 (Academic,
        New York, London 1971) p. 214

2.19   M. Born, K. Huang: In [Ref. 8.17, p. 129]

2.20   M. Born: *Atomtheorie des festen Zustandes*, Nr. 4 (Teubner, Leipzig 1923)

2.21   M. Born: In *Handbuch der Physik*, Vol. XXIV, ed. by H. Geiger and K. Scheel (Springer, Berlin 1933) p. 623

2.22   C. Kittel: *Introduction to Solid State Physics* (Wiley, New York 1967) p. 381

2.23   B. Danovan, J.F. Angress: *Lattice Vibrations* (Chapman and Hall, London 1971) p. 107

2.24   G. Arlt, P. Quadflieg: Phys. Stat. Sol. **25**, 323 (1968)

2.25   W.A. Harrison: Phys. Rev. B **10**, 767 (1974)

2.26   R.M. Martin: Phys. Rev. B **5**, 1607 (1972)

2.27   G.R. Zeller: Phys. Stat. Sol. (b) **65**, 521 (1974)

2.28   K. Hübner: Phys. Stat. Sol. (b) **57**, 627 (1973)

2.29   K. Hübner: Phys. Stat. Sol. (b) **68**, 223 (1975)

2.30   M. Tsuboi: J. Chem. Phys. **40**, 1326 (1964)

2.31   L. Merten: Z. Naturforsch. **17a**, 174 (1962)

2.32   L. Merten: Z. Naturforsch. **17a**, 216 (1962)

2.33   D. Berlincourt, H. Jaffe, L.R. Shiozawa: Phys. Rev. **129**, 1009 (1963)

2.34   K.B. Tolpygo: Sov. Phys. Solid State **2**, 2367 (1961)

2.35   P. Lawaetz: Phys. Stat. Sol. (b) **57**, 535 (1973)

2.36   L.A. Feldkamp, D.K. Steinmann, N. Vagelatos, J.S. King, G. Venkataraman: J. Phys. Chem. Solids **32**, 1573 (1971)

2.37   J.L. Birman: Phys. Rev. **111**, 1510 (1958)

2.38   J.C. Phillips, J.A. Van Vechten: Phys. Rev. Lett. **23**, 1115 (1969)

Chapter 3

3.1   P. Brüesch: *Phonons: Theory and Experiments I*, Springer Ser. Solid-State Sci., Vol. 34 (Springer, Berlin, Heidelberg 1982)

3.2   M.E. Lines, A.M. Glass: *Principles and Applications of Ferroelectrics and Related Materials* (Clarendon, Oxford 1977);
J.C. Burfoot, G.W. Taylor: *Polar Dielectrics and their Applications* (MacMillan, London 1979)

3.3   Landolt-Börnstein: *Ferroelectrics and Related Substances*, Vol. 16 (Springer, Berlin, Heidelberg 1981)

3.4   H. Mueller: Phys. Rev. **57**, 829 (1940); **58**, 565 (1940);
W.P. Mason: Phys. Rev. **72**, 854 (1947)

3.5   A. von Arx, W. Bantle: Helv. Phys. Acta **16**, 211 (1943)

3.6   G. Busch: Helv. Phys. Acta **11**, 269 (1938)

3.7   S. Hoshino, T. Mitsui, F. Jona, R. Repinsky: Phys. Rev. **107**, 1255 (1957)

3.8   C. Yeh, K.F. Casey: Phys. Rev. **144**, 665 (1966)

3.9   W.J. Merz: Phys. Rev. **76**, 1221 (1949)

3.10   E. Fatuzzo, W.J. Merz: *Ferroelectricity* (North-Holland, Amsterdam 1967) p. 11

3.11   W.J. Merz: Phys. Rev. **91**, 513 (1953)

3.12   W. Känzig: In *Solid State Physics*, Vol. 4 (Academic, New York 1957) p. 57

3.13   W. Känzig, R. Meier: Helv. Phys. Acta **21**, 585 (1949)

3.14   W. Jantsch: In *Dynamical Properties of IV–VI Compounds*, Springer Tracts Mod. Phys., Vol. 99 (Springer, Berlin, Heidelberg 1983) p. 1

3.15   L.D. Landau: Phys. Z. Sowjetunion **11**, 26 (1937)

3.16   A.F. Devonshire: Phil. Mag. **40**, 1040 (1949)

3.17   A.F. Devonshire: Phil. Mag. **42**, 1065 (1951)

3.18   A.F. Devonshire: Adv. Phys. **3**, 85 (1954)

3.19   H. Fröhlich: *Theory of Dielectrics* (Clarendon, Oxford 1949)

3.20   W. Cochran: Adv. Phys. **9**, 387 (1960)

3.21   R. Blinc, B. Zeks: *Soft Modes in Ferroelectrics and Antiferroelectrics* (North-Holland, Amsterdam 1974)

3.22   C. Kittel: *Introduction to Solid State Physics* (Wiley, New York 1967)

3.23   C. Kittel: Phys. Rev. **82**, 729 (1951)

3.24   A.A.Maradudin: In *Ferroelectricity* (Proc. Symp. on Ferroelectricity 1966), ed. by E.F. Weller (Elsevier, Amsterdam 1967) p. 72
3.25   A.S. Barker, Jr., M. Tinkham: Phys. Rev. **125**, 1527 (1962)
3.26   J. Petzelt, G.V. Kozlow, A.A. Volkov: Europhys. News **15**, 1 (July 1984)
3.27   N.S. Gillis: Phys. Rev. B **5**, 1925 (1972)
3.28   H. Thomas: In *Structural Phase Transitions and Soft Modes*, ed. by E.J. Samuelsen, E. Anderson, J. Feder (Universitetsforlaget, Oslo 1971) p. 15
3.29   B.D. Silverman: Phys. Rev. **135**A, 1596 (1964)
3.30   E. Pytte: Phys. Rev. **5**, 3758 (1972)
3.31   J.F. Scott: Rev. Mod. Phys. **46**, 83 (1974)
3.32   M.T. Yin, M.L. Cohen: Phys. Rev. B **26**, 3259 and 5668 (1982)
3.33   K. Kune, R.M. Martin: Phys. Rev. Lett. **48**, 406 (1982)
3.34   L.L. Boyer: Ferroelectrics **35**, 83 (1981)
3.35   W. Porod, P. Vogl: In *Physics of Narrow Gap Semiconductors*, ed. by E. Gornik, H. Heinrich, L. Palmetshofer (Springer, Berlin, Heidelberg 1982) p. 247
3.36   W. Porod, P. Vogl, G. Bauer: J. Phys. Soc. Jpn. **49**, Suppl. A, 649 (1980)
3.37   R.P. Lowndes: Phys. Rev. B **6**, 1490 (1972)
3.38   J.C. Slater: Phys. Rev. **78**, 748 (1950)
3.39   V.M. Fridkin: *Ferroelectric Semiconductors* (Consultants Bureau, New York 1980)
3.40   R. Migoni, H. Bilz, D. Bäuerle: Phys. Rev. Lett. **37**, 1155 (1976)
3.41   M. Balkanski, M.K. Teng, M. Massot, H. Bilz: Ferroelectrics **26**, 737 (1980)
3.42   H. Bilz, H. Büttner, A. Bussmann-Holder, W. Kress, U. Schröder: Phys. Rev. Lett. **48**, 264 (1982)
3.43   H. Bilz, A. Bussmann, G. Benedek, H. Büttner, D. Strauch: Ferroelectrics **25**, 339 (1980)
3.44   P. Brüesch, M. Lietz: J. Phys. Chem. Solids **31**, 1137 (1970)
3.45   A. Bussmann-Holder, H. Bilz, P. Vogl: In *Dynamical Properties of IV–VI Compounds*, Springer Tracts Mod. Phys., Vol. 99 (Springer, Berlin, Heidelberg 1983) p. 74
3.46   A. Bussmann-Holder, H. Bilz, D. Bäuerle, D. Wagner: Z. Phys. B **41**, 353 (1981)
3.47   G. Verstraeten: Z. Phys. B **43**, 149 (1981)
3.48   Y. Luspin, J.L. Servoin, F. Gervais: J. Phys. C **13**, 3761 (1980)
3.49   G. Burns: Phys. Lett. **43**A, 271 (1973)
3.50   A. Scalabrin, A.S. Chaves, D.S. Shim, S.P.S. Porto: Phys. Stat. Sol. (b) **79**, 731 (1977)
3.51   C.H. Perry, D.B. Hall: Phys. Rev. Lett. **15**, 700 (1965)
3.52   W. Stirling: J. Phys. C **5**, 2711 (1972)
3.53   R.A. Cowley: Phys. Rev. Lett. **9**, 159 (1962)
3.54   Y. Yamada, G. Shirane: J. Phys. Soc. Jpn. **26**, 396 (1969)
3.55   R.A. Cowley, W.J.L. Buyers, G. Dolling: Solid State Commun. **7**, 181 (1969)
3.56   G.V. Kozlov, A.A. Volkov, J.F. Scott, G.E. Feldkamp: Phys. Rev. B **28**, 255 (1983)
3.57   E. Sandvold, E. Courtens: Phys. Rev. B **27**, 5660 (1983)
3.58   G. Sorge, U. Straube: Phys. Stat. Sol. (a) **51**, 117 (1979)
3.59   A. Levstik, C. Filipic, R. Blinc: Solid State Commun. **18**, 1231 (1976)
3.60   H. Barrett: Phys. Rev. **86**, 118 (1952)
3.61   G.A. Samara: Phys. Rev. Lett. **27**, 103 (1971); Ferroelectrics **20**, 87 (1978) and **22**, 925 (1979)
3.62   D. Rytz, U.T. Höchli, H. Bilz: Phys. Rev. B **22**, 359 (1980)
3.63   K.K. Kobayashi: J. Phys. Soc. Jpn. **24**, 497 (1968)
3.64   T. Schneider, H. Beck, E. Stoll: Phys. Rev. B **13**, 1123 (1976); R. Oppermann, H. Thomas: Z. Phys. B **22**, 387 (1975)
3.65   Recently, it has been suggested that Ta substitution by Nb leads to frequency-dependent susceptibilities reminiscent of an electric-dipole glass [G. Samara: Phys. Rev. Lett. **53**, 298 (1984)]. This interpretation is currently being disputed; W. Kleemann, F.J. Schäfer, D. Rytz: Phys. Rev. Lett. **54**, 2038 (1985)
3.66   D. Moy, R.C. Potter, A.C. Anderson: J. Low Temp. Phys. **52**, 115 (1983), and references therein

3.67 J.J.v.d. Klink, D. Rytz, F. Borsa, U.T. Höchli: Phys. Rev. B **27**, 89 (1983), and references therein
3.68 R.L. Prater, L.L. Chase, L.A. Boatner: Phys. Rev. B **23**, 5904 (1981)
3.69 R. Shuker, W. Gammon: Phys. Rev. Lett. **25**, 222 (1970)
3.70 U.T. Höchli: Phys. Rev. Lett. **48**, 1494 (1982)
3.71 J.A. Mydosh: In *Disordered Systems and Localization*, Lecture Notes Phys., Vol. 149 (Springer, Heidelberg, Berlin 1981)
3.72 R.L. Prater, L.L. Chase, L.A. Boatner: Solid State Commun. **40**, 697 (1981); M. Dubus, B. Daudin, B. Salce, L.A. Boatner: Solit State Commun. **55**, 759 (1985); S.R. Andrews: J. Phys. C **18**, 1357 (1985)
3.73 D. Sherrington, S. Kirkpatrick: Phys. Rev. Lett. **35**, 1792 (1975)
3.74 H. Sompolinsky, A. Zippelius: Phys. Rev. B **23**, 6860 (1982)
3.75 J.D. Reger, K. Binder: Z. Phys. B **60**, 137 (1985)
3.76 A.S. Barker: In [Ref. 3.26, p. 213]
3.77 R. Morf, T. Schneider, E. Stoll: Phys. Rev. B **16**, 462 (1977)
3.78 P. Brüesch: *Phonons: Theory and Experiments II*, Springer Ser. Solid-State Sci., Vol. 65 (Springer, Berlin, Heidelberg 1986)

Chapter 4

4.1 P. Brüesch: *Phonons. Theory and Experiments I*, Springer Ser. Solid-State Sci., Vol. 34 (Springer, Berlin, Heidelberg 1982)
4.2 P. Debye: In *Vorträge über die kinetische Theorie der Materie und Elektrizität* (Teubner, Leipzig 1914); R. Peierls: Ann. Physik **3**, 1055 (1929)
4.3 C. Herring: Phys. Rev. **95**, 954 (1954)
4.4 J. Callaway: Phys. Rev. **113**, 1046 (1959)
4.5 R.E. Nettleton: Phys. Rev. **132**, 2032 (1963)
4.6 M.G. Holland: Phys. Rev. **132**, 2461 (1963)
4.7 P.G. Klemens: In *Solid State Physics*, Vol. 7 (Academic, New York 1958) p. 1
4.8 P. Erdös: Phys. Rev. **138**, A 1200 (1965)
4.9 P. Carruthers: Rev. Mod. Phys. **33**, 92 (1961)
4.10 Ph.D. Thacher: Phys. Rev. **156**, 975 (1967)
4.11 G.A. Slack: Phys. Rev. **105**, 832 (1957)
4.12 R.J. von Gutfeld, A.H. Nethercot, Jr.: Phys. Rev. Lett. **17**, 868 (1966)
4.13 F. Righini, A. Cezairliyan: High Temp.-High Press. **5**, 481 (1973)
4.14 W. Ludwig: In *Phonons and Phonon Interactions*, ed. by T.A. Bak (Benjamin, New York 1964) p. 23
4.15 J.A. Reissland: *The Physics of Phonons* (Wiley, London 1973) p. 166
4.16 J. Callaway: *Quantum Theory of the Solid State*, Pt. A (Academic, New York 1974) p. 55
4.17 R. Berman, E.L. Foster, J.M. Ziman: Proc. Roy. Soc. A **231**, 130 (1955)
4.18 P.G. Klemens: Proc. Phys. Soc. (London) A **68**, 1113 (1955)
4.19 H.E. Jackson, C.T. Walker: Phys. Rev B **3**, 1428 (1971)
4.20 J. Krüger: Phys. Chem. Glasses **13**, 9 (1972)
4.21 R.B. Stephens: Phys. Rev. B **8**, 2896 (1973)
4.22 J.C. Lasjaunias, A. Ravex, M. Vandorpe, S. Hunklinger: Solid State Commun. **17**, 1045 (1975)
4.23 S. Hunklinger, W. Arnold: In *Physical Acoustics* **12**, 155 (Academic, New York 1976)
4.24 R. Berman, F.E. Simons, P.G. Klemens, T.M. Fry: Nature **166**, 864 (1950)
4.25 J.M. Ziman: *Electrons and Phonons* (Oxford U. Press, London 1967) p. 334
4.26 R. Berman, D.K.C. MacDonald: Proc. Roy. Soc. (London) A **211**, 122 (1952)
4.27 L. Halborn: Ann. Physik **59**, 145 (1919)
4.28 W.J. de Haas, J.H. de Boer, G.J. van den Berg: Physica **1**, 1115 (1933/34)
4.29 F. Reif: *Statistical Physics*, Berkeley Physics Course, Vol. 5 (McGraw-Hill, New York 1967) p. 234

4.30 C.C. Ackerman, R.A. Guyer: Ann. Phys. **50**, 128 (1968)
4.31 H.E. Jackson, C.T. Walker, T.F. McNelly: Phys. Rev. Lett. **25**, 26 (1970)
4.32 V. Narayanamurti, R.C. Dynes: Phys. Rev. Lett. **28**, 1461 (1972)
4.33 V. Peshkov: J. Phys. (USSR) **8**, 381 (1944)
4.34 C.C. Ackerman, B. Bertman, H.A. Fairbank, R.A. Guyer: Phys. Rev. Lett. **16**, 789 (1966)
4.35 R.A. Guyer, J.A. Krumhansl: Phys. Rev. **148**, 766 (1966); Phys. Rev. **148**, 778 (1966)
4.36 R.A. Cowley: Proc. Phys. Soc. **90**, 1127 (1967)
4.37 H. Beck: In *Dynamical Properties of Solids*, Vol. 2, ed. by G.K. Horton, A.A. Maradudin (North-Holland, Amsterdam 1975) p. 207; this is an excellent review of the subject
4.38 C. Herring: Phys. Rev. **95**, 954 (1954)
4.39 P. Erdös, S.B. Haley: Phys. Rev. **184**, 951 (1969)
4.40 P. Erdös, S.B. Haley: Phys. Rev. B **4**, 669 (1971)
4.41 F.L. Vook: Phys. Rev. **149**, 631 (1966)
4.42 J.A. Sussmann, A. Thellung: Proc. Phys. Soc. **81**, 1122 (1963)
4.43 R.N. Gurzhi: Zh. eksper. teor. Fiz. **46**, 719 (1964); Sov. Phys.-J. exp. theor. Phys. **19**, 490 (1964); Fiz. tverd. Tela **7**, 3515 (1965); Sov. Phys.-Solid State **7**, 2838 (1966)
4.44 A. Thellung: In *Magnetism in Metals and Metallic Compounds*, ed. by J.T. Lopuszanski, A. Pekalski, and J. Przystawa (Plenum, New York 1976) p. 527
4.45 H. Beck, P.F. Meier, A. Thellung: Phys. Stat. Sol. (a) **24**, 11 (1974)
4.46 M.N. Wybourne, B.J. Kiff, D.N. Batchelder, D. Greig, M. Sahota: J. Phys. C **18**, 309 (1985)
4.47 M.N. Wybourne, B.J. Kiff: Phys. Rev. Lett. **53**, 580 (1984)
4.48 M.N. Wybourne, D.J. Jefferies, L.J. Challis, A.A. Ghazi: Rev. Sci. Instrum. **50**, 1634 (1979)
4.49 θ_D is estimated from $\hbar\omega_D = k_B\theta_D$ and using [Ref. 4.1, Eq. (3.93)]. The mean velocity v is calculated from $3v^{-3} = v_L^{-3} + 2v_T^{-3}$, where v_L and v_T are calculated from the elastic constants for waves propagating along (100) [Ref. 4.1, Problem 3.8.5]. The elastic constants are taken from W.J.L. Buyers: Phys. Rev. **153**, 923 (1967)

Chapter 5

5.1 I.F. Shchegolev: Phys. Stat. Sol. (a) **12**, 9 (1972)
5.2 H.R. Zeller: In *Festkörperprobleme* **13**, 31 (Pergamon/Vieweg, Braunschweig 1973)
5.3 W. Gläser: In *Festkörperprobleme* **14**, 205 (Pergamon/Vieweg, Braunschweig 1974)
5.4 P. Pincus: In *Low Dimensional Cooperative Phenomena*, ed. by H.J. Keller (Plenum, New York 1974) p. 1
5.5 H.R. Zeller, S. Strässler: Comments Sol. State Phys. **7**, 17 (1975)
5.6 H.G. Schuster: In *One-Dimensional Conductors*, Lecture Notes Phys., Vol. 34 (Springer, Berlin, Heidelberg 1975) p. 1
5.7 G.A. Toombs: Phys. Rep. **40**, 181 (1978)
5.8 A.J. Heeger: Comments Sol. State Phys. **9**, 65 (1979); **10**, 53 (1981); **10**, 133 (1982)
5.9 G. Grüner: Comments Sol. State Phys. **10**, 183 (1983)
5.10 R.E. Peierls: Ann. Physik **4**, 121 (1930)
5.11 R.E. Peierls: *Quantum Theory of Solids* (Oxford U. Press, London 1955) p. 108
5.12 W.A. Little: In [Ref. 5.4, p. 35]
5.13 H. Fröhlich: Proc. Roy. Soc. A **223**, 296 (1954)
5.14 A. Heeger, A.F. Garito: In [Ref. 5.4, p. 89] and [Ref. 5.6, p. 151]
5.15 *Conducteurs et Supraconducteurs Synthétiques à basses Dimensions*, J. Physique **44**, C3, Suppl. 6 (1983) (Conférence Internationale sur la Physique et la Chimie des Polymères Conducteurs, 11–15 Déc. 1982)

5.16   K. Krogmann: Angew. Chem. Int. Ed. Engl. **8**, 35 (1969); in [Ref. 5.4, p. 277]

5.17   M.J. Minot, J.H. Perlstein: Phys. Rev. Lett. **26**, 371 (1971)

5.18   W. Kohn: Phys. Rev. Lett. **2**, 393 (1959)

5.19   P. Brüesch: *Phonons: Theory and Experiments I*, Springer Ser. Solid-State Sci., Vol. 34 (Springer, Berlin, Heidelberg 1982)

5.20   P.A. Lee, T.M. Rice, P.W. Andersen: Solid State Commun. **14**, 703 (1974); J. Bardeen: Solid State Commun. **13**, 357 (1973)

5.21   O. Madelung: *Introduction to Solid State Theory*, Springer Ser. Solid-State Sci., Vol. 2 (Springer, Berlin, Heidelberg 1978) p. 22

5.22   L.I. Schiff: *Quantum Mechanics*, 3rd ed. (McGraw-Hill, New York 1968) p. 167

5.23   M.J. Rice, S. Strässler: Solid State Commun. **13**, 125 (1973)

5.24   P.A. Lee, T.M. Rice, P.W. Anderson: Phys. Rev. Lett. **31**, 462 (1973)

5.25   L. Pietronero, S. Strässler, G.A. Toombs: Phys. Rev. B **12**, 5213 (1975)

5.26   M.J. Rice, S. Strässler: Solid State Commun. **13**, 1931 (1973)

5.27   W. Rüegg, D. Kuse, H.R. Zeller: Phys. Rev. B **8**, 952 (1973)

5.28   H.R. Zeller, A. Beck: J. Phys. Chem. Solids **35**, 77 (1974)

5.29   R. Comes, M. Lambert, H. Launois, H.R. Zeller: Phys. Rev. B **8**, 571 (1973)

5.30   P. Brüesch: *Phonons: Theory and Experiments II*, Springer Ser. Solid-State Sci., Vol. 65 (Springer, Berlin, Heidelberg 1986)

5.31   H.R. Zeller, P. Brüesch: Phys. Stat. Sol. (b) **65**, 537 (1974)

5.32   R.P. Messmer, D.R. Salahub: Phys. Rev. Lett. **35**, 533 (1975)

5.33   H. Yersin, G. Gliemann, U. Rössler: Sol. State Commun. **21**, 915 (1975)

5.34   B. Renker, H. Rietschel, L. Pintschovius, W. Gläser, P. Brüesch, D. Kuse, M.J. Rice: Phys. Rev. Lett. **30**, 1144 (1973)

5.35   R. Comes, B. Renker, L. Pintschovius, R. Currat, W. Gläser, G. Scheiber: Phys. Stat. Sol. (b) **71**, 171 (1975)

5.36   M.R. Rice, C.B. Duke, N.O. Lipari: Sol. State Commun. **17**, 1089 (1975)

5.37   J.W. Lynn, M. Iizumi, G. Shirane, S.A. Werner, R.B. Saillant: Phys. Rev. B **12**, 1154 (1975)

5.38   K. Carneiro, G. Shirane, S.A. Werner, S. Kaiser: Phys. Rev. B **13**, 4258 (1976); G. Shirane: Ferroelectrics **16**, 7 (1977)

5.39   P. Brüesch, H.R. Zeller: Solid State Commun. **14**, 1037 (1974)

5.40   P. Brüesch, S. Strässler, H.R. Zeller: Phys. Rev. B **12**, 219 (1975)

5.41   M.J. Rice, S. Strässler: Solid State Commun. **13**, 1389 (1973)

5.42   P. Brüesch: Solid State Commun. **13**, 13 (1973)

5.43   L.S. Agroskin, R.M. Vlasova, A.I. Gutman, R.N. Lynbovskaya, G.V. Papayan, L.P. Rautian, L.D. Rozenshtein: Sov. Phys. Solid State **15**, 1189 (1973)

5.44   P. Brüesch: In [Ref. 5.6, pp. 194–243]

5.45   R.C. Jaklevic, R.B. Saillant: Solid State Commun. **15**, 307 (1974)

5.46   E.F. Steigmeier, R. Loudon, G. Harbeke, H. Auderset, G. Scheiber: Solid State Commun. **17**, 1447 (1975)

5.47   L. Van Hove: Physica **16**, 137 (1950)

5.48   L.D. Landau, E.M. Lifschitz: *Statistical-Physics* (Pergamon, London 1959) p. 482

5 49   M. Blunck, H.G. Reik: Z. Physik B **32**, 147 (1979)

5.50   D. Allender, J.W. Bray, J. Bardeen: Phys. Rev. B **9**, 119 (1974)

5.51   S. Strässler, G.A. Toombs: Phys. Letters A **46**, 321 (1974)

5.52   H. Fukuyama, T.M. Rice, C.M. Varma: Phys. Rev. Lett. **33**, 305 (1974)

5.53   A. Luther, V.J. Emery: Phys. Rev. Lett. **33**, 589 (1974)

5.54   H.R. Zeller: In *Low-Dimensional Cooperative Phenomena*, ed. by H.J. Keller (Plenum, New York 1975) p. 215

5.55   A. Brau, P. Brüesch, J.P. Farges, W. Hinz, D. Kuse: Phys. Stat. Sol. (b) **62**, 615 (1974)

5.56   M.J. Rice, C.B. Duke, N.O. Lipari: Sol. State Commun. **17**, 1089 (1975)

5.57   E.F. Steigmeier, H. Auderset, D. Baeriswyl, M. Almeida, K. Carneiro: J. Physique **44**, C3-1445 (1983)

5.58   M. Apostol, I. Baldea: Solid State Commun. **53**, 687 (1985)

Chapter 6

6.1   R.J. Elliott: In *Phonons in Perfect Lattices and in Lattices with Point Imperfections*, ed. by R.W.H. Stevenson (Oliver and Boyd, Edinburgh 1966) Chap. 14

6.2   A.A. Maradudin: *Solid State Physics* **18**, 273 (Academic, New York 1966); **19**, 1 (1966)

6.3   J. Hori: *Spectral Properties of Disordered Chains and Lattices* (Pergamon, New York 1968)

6.4   A.A. Maradudin, E.W. Montroll, G.H. Weiss, I.P. Ipatova: *Theory of Lattice Dynamics in the Harmonic Approximation*, Solid State Physics, Suppl. 3 (Academic, New York 1971)

6.5   R.J. Elliott, J.A. Krumhansl, P.L. Leath: Rev. Mod. Phys. **46**, 465 (1974)

6.6   A.S. Barker, A.J. Sievers: Rev. Mod. Phys. **47**, Suppl. 2, S1 (1975)

6.7   D.W. Taylor: In *Dynamical Properties of Solids*, Vol. 2, ed. by G.K. Horton and A.A. Maradudin (North-Holland, Amsterdam 1975) Chap. 5

6.8   R.J. Elliott, P.L. Leath: In [Ref. 6.7, Chap. 6]

6.9   H. Böttger: *Principles of the Theory of Lattice Dynamics* (Akademie-Verlag, Berlin 1983)

6.10   D.N. Zubarev: Usp. Fiz. Nauk **71**, 71 (1960) [Sov. Phys.-Usp. **3**, 320 (1960)]

6.11   F.J. Dyson, Phys. Rev. **92**, 1331 (1953)

6.12   S. Alexander, J. Bernasconi, W.R. Schneider, R. Orbach: Rev. Mod. Phys. **53**, 175 (1981)

6.13   P. Dean: Proc. R. Soc. (Lond.) A **260**, 263 (1961)

6.14   D.N. Payton, III, W.M. Visscher: Phys. Rev. **154**, 802 (1967)

6.15   P. Dean: Rev. Mod. Phys. **44**, 127 (1972)

6.16   D.W. Taylor: Phys. Rev. **156**, 1017 (1967)

6.17   J.E. Hirsch: Phys. Rev. B **18**, 3976 (1978)

6.18   C.W. Myles, J.D. Dow: Phys. Rev. B **19**, 4939 (1979)

6.19   C.E.T. Gonçalves da Silva, B Koiller: Solid State Commun. **40**, 215 (1981)

6.20   J.-M. Langlois, A.-M.S. Tremblay, B.W. Southern: Phys. Rev. B **28**, 218 (1983)

6.21   M.O. Robbins, B. Koiller: Phys. Rev. B **27**, 7703 (1983)

6.22   J. Bernasconi (unpublished)

6.23   B. Koiller, M.O. Robbins, M.A. Davidovich, C.E.T. Gonçalves da Silva: Solid State Commun. **45**, 955 (1983)

6.24   G.P. Montgomery, Jr., M.V. Klein, B.N. Ganguly, R.F. Wood: Phys. Rev. B **6**, 4047 (1972)

6.25   Y. Tsunoda, N. Kunitomi, N. Wakabayashi, R.M. Nicklow, H.G. Smith: Phys. Rev. B **19**, 2876 (1979)

Chapter 7

7.1   S. Geller (ed.): *Solid Electrolytes*, Topics Appl. Phys., Vol. 21 (Springer, Berlin, Heidelberg 1977)

7.2   S. Chandra: *Superionic Solids* (North-Holland, Amsterdam 1981)

7.3   *Fast Ionic Transport in Solids*, ed. by J.B. Bates and G.C. Famington (Proc. Intern. Conf. on Fast Ionic Transport in Solids, Gatlinburg, TE (North-Holland, Amsterdam 1981)

7.4   W. Dieterich, P. Fulde, I. Peschel: Adv. Phys. **29**, 527 (1980)

7.5a   W. Dieterich: Phys. Bl. **36**, 354 (1980)

7.5b   W. Dieterich: In *Festkörperprobleme* **21**, 325 (Vieweg, Braunschweig 1981)

7.6   C.R.A. Catlow: Solid State Ionics **8**, 89 (1983)

7.7   H. Schulz: Ann. Rev. Mater. Sci. **12**, 351 (1982)

7.8   C.R.A. Catlow: Comments Solid State Phys. **9**, 157 (1980)

7.9   M.B. Salamon (ed.): *Physics of Superionic Conductors*, Topics Cur. Phys., Vol. 15 (Springer, Berlin, Heidelberg 1979)

7.10   J.B. Boyce, B.A. Huberman: Phys. Repts. **51**, 189 (1979)

7.11   W. Hayes: Contemp. Phys. **19**, 469 (1978)

7.12   K. Funke: Semiconductors and Insulators **3**, 351 (1978)

7.13   H.R. Zeller: Phys. Bl. **33**, 612 (1977)
7.14   K. Shahi: Phys. Stat. Sol. (a) **41**, 11 (1977);
        M. O'Keeffe: Comments Sol. State Phys. **7**, 163 (1977)
7.15   H.U. Beyeler, P. Brüesch: In *Current Topics in Materials Science* **2**, 765 (North-Holland, Amsterdam 1976)
7.16   R.D. Armstrong, R.S. Bulmer, T. Dickinson: J. Solid State Chem. **8**, 219 (1973)
7 17   G.D. Mahan, W.L. Roth (eds.): *Superionic Conductors* (Plenum, New York 1976)
7.18   P. Brüesch: *Phonons: Theory and Experiments I*, Springer Ser. Solid-State Sci., Vol. 34 (Springer, Berlin, Heidelberg 1982)
7.19   L.W. Strock: Z. Physik Chem. B **25**, 411 (1934); B **31**, 132 (1936)
7.20   A.B. Lidiard: "Ionic Conductivity", in *Electrical Conductivity* II, Encyclopedia of Physics, Vol. 20, ed. by S. Flügge (Springer, Berlin, Göttingen 1957) p. 246
7.21   T. Geisel: In [Ref. 7.9, p. 201]
7.22   M.J. Rice, W.L. Roth: J. Solid State Chem. **4**, 294 (1972)
7.23   H.U. Beyeler, P. Brüesch, L. Pietronero, W.R. Schneider, S. Strässler, H.R. Zeller: In [Ref. 7.9, p. 77]
7.24   T. Geisel: In [Ref. 7.9, p. 206]
7.25   F.G. Mahan, W.J. Pardee: Phys. Lett. **49** A, 325 (1974)
7.26   W.J. Pardee, G.D. Mahan: J. Solid State Chem. **15**, 310 (1975)
7.27   H. Hinkelmann, B.A. Huberman: Solid State Commun. **19**, 365 (1976)
7.28   H. Hinkelmann: Solid State Commun. **21**, 975 (1977)
7.29   H.R. Zeller, P. Brüesch, L. Pietronero, S. Strässler: In [Ref. 7.19, p. 201]
7.30   P. Brüesch, L. Pietronero, S. Strässler, H.R. Zeller: Phys. Rev. B **15**, 4631 (1977)
7.31   C.P. Flynn, A.M. Stoneham: Phys. Rev. B **1**, 3966 (1970)
7.32   P. Fulde, L. Pieteronero, W.R. Schneider, S. Strässler: Phys. Rev. Lett. **35**, 1776 (1975)
7.33   P. Brüesch, S. Strässler, H.R. Zeller: Phys. Stat. Sol. (a) **31**, 217 (1975)
7.34   P. Brüesch, L. Pietronero, H.R. Zeller: J. Phys. C **9**, 3977 (1976)
7.35   W. Dieterich, I. Peschel, W.R. Schneider: Z. Physik B **27**, 177 (1977)
7.36   H. Risken, H.D. Vollmer: Z. Physik B **33**, 297 (1979)
7.37   P. Brüesch, L. Pietronero, S. Strässler, H.R. Zeller: Electrochimica Acta **22**, 717 (1977)
7.38   T. Geisel: Phys. Rev. B **20**, 4294 (1979)
7.39   B.A. Huberman, R.M. Martin: Phys. Rev. B **13**, 1498 (1976)
7.40   K.R. Subbaswami: Solid State Commun. **19**, 1157 (1976); **21**, 371 (1977)
7.41   H. Hayashi, M. Kobayashi, I. Yokota: Solid State Commun. **31**, 847 (1979)
7.42   J. Jäckle: Z. Physik B **30**, 255 (1978)
7.43   R. Zeyher: Z. Physik B **31**, 127 (1978)
7.44   R. Zeyher: Solid State Commun. **36**, 33 (1980)
7.45   N. Mizoguchi, H. Hayashi, M. Kobayashi, I. Yokota: J. Phys. Soc. Jpn. **50**, 2043 (1981)
7.46   G. Jacucci, A. Rahman: J. Chem. Phys. **69**, 4117 (1978)
7.47   W. Schommers: Phys. Rev. Lett. **38**, 1536 (1977)
7.48   P. Vashishta, A. Rahman: Phys. Rev. Lett. **40**, 1337 (1978)
7.49   A. Fukumoto, A. Ueda, Y. Hiwatari: Solid State Ionics **3/4**, 115 (1981)
7.50   M.J. Gillan: Solid State Ionics **9/10**, 755 (1983)
7.51   R. Kubo: Rep. Prog. Phys. **29**, 225 (1966)
7.52   W.R. Schneider: Z. Physik B **24**, 135 (1976)
7.53   M. Bixon, R. Zwanzig: J. Stat. Phys. **3**, 245 (1971)
7.54   D.H. Menzel (ed.): *Fundamental Formulas of Physics*, Vol. 1 (Dover, New York 1960) p. 59
7.55   S. Chandrasekhar: Rev. Mod. Phys. **15**, 68 (1943)
7.56   C.A. Wert: Phys. Rev. **79**, 601 (1950)
7.57   Y.M. Ivanchenko, L.A. Zil'berman: Zh. Eksp. Teor. Fiz. **55**, 2395 (1968) [Sov. Phys.-JETP **28**, 1272 (1969)]
7.58   H. Mori: Progr. theor. Phys. (Kyoto) 33, 423 (1965)
7.59   P. Brüesch: *Phonons: Theory and Experiments II*, Springer Ser. Solid-State Sci., Vol. 65 (Springer, Berlin, Heidelberg 1986)

7.60   R. Shuker, R.W. Gammon: Phys. Rev. Lett. **25**, 222 (1970)
7.61   F.L. Galeener, P N. Sen: Phys. Rev. B **17**, 1928 (1978)
7.62   K. Funke, A Gacs, H.J. Schneider, S.M. Ansari, N. Martinkat, H. Roemer, H.G. Unruh: Solid State Ionics **11**, 247 (1983);
H. Roemer, D. Schwarz, H.G. Unruh, G. Luther, K. Funke: Solid State Ionics **11**, 253 (1983)
7.63   P. Brüesch, W. Bührer, H.J.M. Smeets: Phys. Rev. B **22**, 970 (1980)
7.64   P. Brüesch, H.U. Beyeler, W. Bührer: In *Proceedings of the International Conference on Lattice Dynamics*, ed. by M. Balkanski (Flammarion, Paris 1977) p. 527
7.65   R. Alban, G. Burns: Phys. Rev. B **16**, 3746 (1977)
7.66   G. Burns, R. Alben, F.H. Dacol, M.W. Shafer: Phys. Rev. B **20**, 638 (1979)
7.67   H. Böttger: *Principles of Lattice Dynamics* (Akademie-Verlag, Berlin 1983) p. 192
7.68   A.S. Barker, Jr., R. Loudon: Rev. Mod. Phys. **44**, 18 (1972)
7.69   R.H. Stolen: Phys. Chem. Glasses **11**, 83 (1970)
7.70   G. Burns, F.H. Dacol, M.W. Shafer: Phys. Rev. B **16**, 1416 (1977)
7.71   W. Bührer, R.M. Nickow, P. Brüesch: Phys. Rev. B **17**, 3362 (1978)
7.72   P. Brüesch, H.U. Beyeler, S. Strässler: Phys. Rev. **25**, 541 (1982)
7.73   B. Gras, K. Funke: Solid State Ionics **2**, 341 (1981)
7.74   G. Eckold, K. Funke: Z. Naturforsch. **28a**, 1042 (1973)
7.75   M.J. Delaney, S. Ushioda: Solid State Commun. **19**, 297 (1976)
7.76   D. Gallagher, M.V. Klein: J. Phys. C **9**, L687 (1976)
7.77   P. Brüesch, T. Hibma, W. Bührer: Phys. Rev. B **27**, 5052 (1983)
7.78   P. Brüesch, W. Bührer, R.S. Perkins: J. Phys. C **10**, 4023 (1977)
7.79   G. Burns, F.H. Dacol, M.W. Shafer, R. Alben: Solid State Commun. **24**, 753 (1977)
7.80   G. Burns, R. Alben, F.H. Dacol, M.W. Shafer: Phys. Rev. B **20**, 638 (1979)
7.81   K. Funke: In [Ref. 7.17, p. 183]
7.82   R.J. Elliott, W. Hayes, W.G. Kleppmann, A.J. Rushworth, J.F. Ryan: Proc. R. Soc. Lond. A **360**, 317 (1978)
7.83   W. Hayes, A.M. Stoneham: *Crystals with the Fluorite Structure*, ed. by W. Hayes (Clarendon, Oxford 1974) Chap. 2
7.84   S.J. Allen, Jr., J.P. Remeika: Phys. Rev. Lett. **33**, 1478 (1974)
7.85   S.J. Allen, Jr., A.S. Cooper, F. DeRosa, J.P. Remeika, S.K. Ulasi: Phys. Rev. B **17**, 4031 (1978)
7.86   Ph. Colomban, G. Lucazeau: J. Chem. Phys. **72**, 1213 (1980)
7.87   Ph. Colomban, R. Mercier, G. Lucazeau: J. Chem. Phys. **75**, 1388 (1981)
7.88   C.P. Flynn: *Point Defects and Diffusion* (Clarendon, Oxford 1972) p. 327
7.89   C.P. Flynn: Phys. Rev. **171**, 682 (1968)
7.90   S.A. Rice: Phys. Rev. **112**, 804 (1958)
7.91   M. Kac: Am. J. Math. **65**, 609 (1943)
7.92   N.B. Slater: *Theory of Unimolecular Reactions* (Cornell U. Press, Ithaca, NY 1959)
7.93   D. Feit: Phys. Rev. B **3**, 1223 (1971)
7.94   T.M. Haridasan, J. Govindarajan, M.A. Nerenberg, P.W.M. Jacobs: J. Phys. C **13**, 3107 (1980)
7.95   N.B. Slater: Proc. Roy. Soc. (London) A **194**, 112 (1948)
7.96   M. Dixon, M.J. Gillan: J. Phys. C **11**, L165 (1978)
7.97   A.B. Walker, M. Dixon, M.J. Gillan: J. Phys. C **15**, 4061 (1982)
7.98   M. Parinello, A. Rahman, P. Vashishta: Phys. Rev. Lett. **50**, 1073 (1983)
7.99   M.H. Dickens, W. Hayes, M.T. Hutchings, C. Smith: J. Phys. C **15**, 4043 (1982)
7.100   M.T. Hutchings, K. Clausen, M.H. Dickens, W. Hayes, J.K. Kiems, P.G. Schnabel, C. Smith: J. Phys. C **17**, 3903 (1984)
7.101   S. Brawer: Phys. Rev. **10**, 3287 (1974)
7.102   A. Höch: Quasielastische Streuung kalter Neutronen an Einkristallen der Hochtemperaturphasen von AgI und Ag$_2$Se, Dissertation, University of Hannover (1983)
7.103   H.U. Beyeler, L. Pietronero, S. Strässler: Phys. Rev. B **22**, 2988 (1980)
7.104   H.E. Roman, A. Bunde, W. Dieterich: Phys. Rev. B **34**, 3439 (1986)

7.105 S. Glasstone, K.J. Laidler, H. Eyring: *The Theory of Rate Processes* (McGraw Hill, New York 1941) pp. 15, 189
7.106 J.M. Thomas, W.J. Thomas: *Introduction to the Principles of Heterogeneous Catalysis* (Academic, New York, London 1967) p. 51

Chapter 8

8.1 R.M.J. Cotterill: Phys. Bull. **32**, 285 (1981)
8.2 A.R. Ubbelohde: *The Molten State of Matter* (Wiley, New York 1978)
8.3 J.E. Lennard-Jones, A.F. Devonshire: Proc. R. Soc. A **170**, 464 (1939)
8.4 S.M. Stishov, I.N. Makarenko, V.A. Ivanov, A.M. Nikolaenko: Phys. Lett. A **45**, 18 (1973)
8.5 F. Lindemann: Phys. Z. **11**, 609 (1910)
8.6 J.A. Reissland: *The Physics of Phonons* (Wiley, London 1973) p. 132
8.7 D. Ter Haar (ed.): *Collected Papers of L.D. Landau* (Pergamon, London 1965) p. 193
8.8 T.V. Ramakrishnan, M. Yussouff: Phys. Rev. B **19**, 2775 (1979)
8.9 R.M.J. Cotterill, E.J. Jensen, W.D. Kristensen: *Anharmonic Lattices, Structural Transitions and Melting*, ed. by T. Riste (Noordhoff, Leiden 1974) p. 405; J.K. Kristensen, R.M.J. Cotterill: *Physics of Non Equilibrium Systems: Fluctuations, Instabilities and Phase Transitions*, ed by T. Riste (Noordhoff, Leiden 1975)
8.10 D. Kuhlmann-Wilsdorf: Phys. Rev. **140**, 1599 (1965)
8.11 D.J. Thouless, A.M. Kosterlitz: J. Phys. C **6**, 1181 (1973); B.I. Halperin, D.R. Nelson: Phys. Rev. Lett. **41**, 121 (1978)
8.12 J.K. Kristensen, R.M.J. Cotterill: Phil. Mag. **36**, 137 (1977)
8.13 P. Brüesch: *Phonons: Theory and Experiments I*, Springer Ser. Solid-State Sci., Vol. 34 (Springer, Berlin, Heidelberg 1982)
8.14 L.L. Boyer: Phys. Rev. Lett. **42**, 584 (1979)
8.15 H. Fukuyama, P.M. Platzman: Solid State Commun. **15**, 677 (1974)
8.16 L. Pietronero, E. Tosatti: Solid State Commun. **32**, 255 (1979)
8.17 J.Q. Broughton, L.W. Woodcock: J. Phys. C **11**, 2743 (1978); J.Q. Broughton, G.H. Gilmer: J. Chem. Phys. **79**, 5119 (1983); Also for a grain boundary a sort of premelting has been observed: G. Ciccotti, M. Guillopé, V. Pontikis: Phys. Rev. B **27**, 5576 (1983)
8.18 S.D. Kevan, D.A. Shirley: Phys. Rev. B **22**, 542 (1980)
8.19 C.S. Jayanthi, E. Tosatti, L. Pietronero: To be published
8.20 P.R. Couchman, W.A. Jesser: Phil. Mag. **35**, 787 (1977); R.M.J. Cotterill: Phil. Mag. **32**, 1283 (1975)
8.21 D.P. Woodruff: *The Solid-Liquid Interface* (Cambridge U. Press, Cambridge 1973)
8.22 P. Brüesch: *Phonons: Theory and Experiments II*, Springer Ser. Solid-State Sci., Vol. 65 (Springer, Berlin, Heidelberg 1986)
8.23 Joost W.M. Frenken, J.F. van der Veen: Phys. Rev. Lett. **54**, 134 (1985)

Subject Index

222

229

237

240

241